BIOPOLITICS

THOMAS LANDON THORSON

UNIVERSITY OF TORONTO

HOLT, RINEHART AND WINSTON, INC.
NEW YORK   CHICAGO   SAN FRANCISCO   ATLANTA
DALLAS   MONTREAL   TORONTO
LONDON   SYDNEY

For Ingrid and Carla

Copyright © 1970 by Holt, Rinehart and Winston, Inc.
All rights reserved.
Library of Congress Catalog Card Number: 77–99107

SBN: 03–083142–3

Printed in the United States of America

1 2 3 4 5 6 7 8 9

Preface

This is a book about politics and about what is involved in understanding politics. In some sense it is about political science and by implication about social science in general. My basic point concerns the implications of a choice between two fundamental conceptual orientations—or, as I call them, paradigms of understanding—when it comes to making sense of politics and political events.

The book is not a survey of all the relevant literature; it is not a history of contemporary political or social thought. Its literary form I would call an argument, although I suppose some might want to use the word polemic.

One of the ways in which books about political science or social science are categorized these days is whether the book is "pro-science" or "anti-science." This is definitely a "pro-science" book but, I think, with something of an unusual twist. American social scientists have over the last half century or so behaved with respect to natural scientists like the members of an underdeveloped culture in contact with a more advanced civilization. They have grasped for the trappings of science and followed the manner of scientists; and in so doing they have beyond much doubt developed some technical skills and an increased understanding of certain subject areas. The time has come, however—and this is my argument—for taking the substantive teachings of science really seriously.

Natural science has a method to teach it is true, but much more importantly it has substantive knowledge to teach. Curiously enough, as they apply to human social behavior, the two are not automatically compatible. This incompatibility is part of what this book is about. Self-consciously adopting what is ordinarily understood as the *method* of science has led to the acceptance of what I call the universal-generalization paradigm of scientific understanding. I, of course, make no claim here about each and every one of the thousands of political scientists now practicing their profession in one form or another. I do, however, want to call critical attention to what it is safe to call a strong tendency among them. The present editor of the *American Political Science Review*—is in my judgment quite correct when he says that, "Most political scientists . . . use the term 'science' to denote a particular method of inquiry . . ." and further that, "The main objective of science is to discover, state, and verify the fundamental 'laws' that govern the universe."[1] Professor Ranney further describes the universal-generalization paradigm of scientific understanding when he says, "Scientists seek generalizations that in the manner of the law of gravitation or the law of the conservation of energy, apply to particular phenomena wherever they may be found, and not merely as they appear in some places. The ultimate goal of science is *systematic* theory—that is, a body of logically consistent and connected statements explaining all aspects of the universe."[2]

These statements are intended as part of an obvious, rudimentary description of what science is—they are part of an introductory text— and what they demonstrate beyond much doubt is that science commonly is defined not substantively but methodologically in accordance with the pursuit of any-time-any-place laws which govern phenomena, laws epitomized by those of nineteenth-century physics. This notion even though it is on the surface only a method has "paradigm force"; it pushes and molds theory and it directs research. I quote now the current president of the American Political Science Association speaking of a unified social theory: "The expectation and hope that it will be possible to develop a common underlying social theory impel research in certain inescapable directions. The most significant of these for our purposes is that it has led to the

---

[1] Austin Ranney, *The Governing of Men* (New York: Holt, Rinehart and Winston, 1966), 2d Ed. Rev., p. 626.
[2] *The Governing of Men*, p. 628.

search for a common unit of analysis that could easily feed into the special subject matters of each of the disciplines. Ideally, the units would be repetitious, ubiquitous, and uniform, molecular rather than molar. In this way they would constitue the particles, as it were, out of which all social behavior is formed and which manifest themselves through different institutions, structures, and processes."[3]

The method of science so understood leads social scientists to search for general laws of behavior and for "uniform" and "ubiquitous" "particles" upon which to base these laws. The substance of science, however, teaches that man is a primate of a special sort, the product of billions of years of tedious natural "trial and error." The *content* of biology, geology, astronomy, palaeontology, and indeed of chemistry and physics themselves give us quite a different angle on human behavior than does the *method* of nineteenth-century physics. The content tells us that behavior has changed and evolved over time but the method tells us we should be seeking constant and unchanging laws about behavior. It is this contrast that I seek here to explore. I ask and try to answer two questions: (1) What are the consequences for political science of adopting the universal-generalization paradigm of scientific understanding? and (2) What kind of consequences would the adoption of an evolutionary-developmental pradigm of scientific understanding be likely to have for political science? I call my first answer a critique and my second a speculation. They are both put argumentatively because, quite frankly, I think an argument needs to be started.

Because much of my discussion involves information and points of view distant from political science and political theory, I have in some cases quoted at greater length than is conventional. Where I thought that someone else said better what I could only have paraphrased, I have not hesitated to present his words. At the same time, however, I have usually resisted the temptation to pile up footnotes and to pay lip service to contrary points of view. The argument I make is controversial at almost every turn and I have no desire to hide the fact. On the other hand I do want the reader to be able to see what I have to say as a whole; and extensive distinction drawing would only obscure that view.

During the making of this book I have been variously associated with the University of Wisconsin, Oxford University, the University of California, Berkeley, the University of the Philippines, Northwestern University, and the University of Toronto. All of these experiences have added in one way or another to what is written here. I want especially to record

---

[3] David Easton, "The Current Meaning of 'Behavioralism'" in James C. Charlesworth, ed., *Contemporary Political Analysis* (New York: The Free Press, 1967), p. 24.

my debt to the John Simon Guggenheim Memorial Foundation for a fellowship in 1962–1963 and to the Rockefeller Foundation and Kenneth W. Thompson in particular for continuing in difficult circumstances to support me after my return from the Philippines in 1966. For listening to my ideas both tolerantly and critically I thank especially Charles W. Anderson of the University of Wisconsin, Douglas W. Rae of Yale University, Wayne A. Kimmel of the Department of Health, Education, and Welfare, and Sondra Thorson,

Books are parts of lives and special thanks go to the following people for helping me with mine, for the most part by just being themselves: Helen Sebsow, now of the United States Information Agency; Nancy Mintz Sills, now of Harper and Row; Lucien A. Gregg, M.D., of the Rockefeller Foundation: Iluminada Panlilio, Florentino Herrera, M.D., Oskar Baguio, and Augustin Cailao of the University of the Philippines; Norman Jacobson of the University of California, Berkeley; and Fred R. von der Mehden, now of Rice University. Fred H. Wilt, now of the Department of Zoology, University of California, Berkeley, and Donald F. Summers, M.D., now of Albert Einstein College of Medicine in New York, goaded me into learning something about biology when we were all undergraduates. James L. McCamy introduced me to the discussions of the Interdisciplinary Committee on the Future of Man at the University of Wisconsin. Herbert Addison of Holt, Rinehart and Winston has been a continuing source of help and encouragement. David Bidney and the late Norwood Russell Hanson were once my teachers and their influence shows here.

Nothing would make me happier than if these pages were to stimulate you to read (if you haven't already) the works of the various authors from whom I have quoted and to whom I have referred. I have found their work fascinating—even those of whom I am critical—and my greatest debt is to them.

T. L. T.

July 1969
LaPorte, Indiana

Contents

History is made by example. Somebody started it. And you are not gonna explain things that happen because there was a real need for it. Because history is made by example, by third rate or even fourth rate people who are unknown, who are buried in unmarked and unvisited graves. To me, always when I see something new coming up in history, see, I'm always certain, you know, somebody was there, somebody we don't know nothing about. And I know from my own life. You take for instance—have you ever been to San Joaquin valley? You have a whole string of towns there. You have Corfat, Modesto, Madeira, Masset. I want to show you how I generalize about history from small experiences that happen to me. See, trivial things. Now, all these towns have the same climate, the same economical conditions, they raise the same crop. But only Modesto has the most beautiful lawns in Christendom. Nowhere in the world will you, are you going to find such beautiful lawns. The whole town is covered with lawns. These people don't seem to be doing nothing but mow the lawn. If the lawn starts to shift, to break up the sidewalk, nobody stops them. They go! The—uh—park commission of the city bred a special ash tree that sheds all its leaves in one week so it won't dirty the lawns. So how come the lawns of Modesto? Corfat hasn't got it, Madeira hasn't got it. Now if I was a sociologist I probably would do research. I don't know what I would do. I didn't know what to do. I had a hunch you know, that what the scientist does is counts. So I counted the churches. Plenty churches, plenty chiropractors. But what the hell connection could there be between churches and chiropractors, and lawns? And then I had a hunch. I went to the cemetery and looked up the oldest graves. The people came from Essex, Essex!! That's where lawns were invented!! And this is the way things change. And so . . .

Eric Hoffer, in a television
interview with Eric Sevareid

Introduction
A Picture of Politics

Politics is many things. It is the activity of dispute over questions petty or grand which arise among men or among groups of men. It is sometimes the settlement of such disputes or perhaps it is merely the attempt at settlement. A proposed definition of "politics," or what is sometimes called "the political," almost inevitably involves the selection of some particular sort of human activity which will serve as a kind of simple picture of "the political." The proposals just made stress bargaining or influence or, perhaps, if properly amplified, power. The picture is of the haggle of customer and street vendor or perhaps the moves and countermoves of parents confronting their children.

It is most likely, nay, even certain, that there is no single picture, however skillfully devised, that will serve as *the proper* picture of politics for all. Indeed, politics can sometimes be a dispute about which picture of politics is correct. The philosopher Ludwig Wittgenstein saw something of great importance when he suggested that the terms and phrases of ordinary language do not have essential meanings to which all varieties of meaning can be reduced or, if not reducible, from which they can be eliminated and condemned as false. He proposed instead the notion of a family of meanings, the members of which have overlapping characteristics

1

but not necessarily any essential ones. Insofar as we receive "politics" from ordinary language and do not assign it a merely prescriptive definition, looking for and disputing its essence is a fruitless task.

Looking carefully at the brothers and sisters and cousins and, so to speak, trying them on for size is, however, by no means without merit. On the contrary, such examination may open the door to greater understanding, which is surely what the study of politics is all about. I propose here no quarrel with those who wish to see politics as bargaining. I want only to suggest that in one of its most important manifestations politics has little in common with the bargain, but is much more nearly akin to certain of the characteristics normally associated with art, religion, philosophy, and science. When politics is concerned with the invention, the promulgation, and the implementation of a plan or a system of social and political order, it is much less like the strategy and tactics of contract bridge or chess or poker than it is like inventing the game itself.

The suggestion was made a moment ago that politics can sometimes be understood on the model of art, religion, philosophy, and science. This perhaps seems a remarkably disparate combination of activities to supply a single model or, as we put it earlier, picture of politics. This, I think, is because we are most often concerned with distinguishing art from science or science from philosophy or philosophy from religion, and we tend therefore to lose sight of their common properties. I would like to argue that these activities, however different they may be in some respects, do have characteristics in common, although again it would be a mistake to propose some *essential* property to which all are reducible. Here the notion of overlapping "family" characteristics is once more helpful. My contention is that art, religion, philosophy, science, *and* politics are all in certain respects *like* one another but none is ever *identical* to another. Thus, politics, which is here the prime object of attention, is sometimes comparable to art, religion, philosophy, and science, is sometimes more like one than another, and yet is always politics and not art, religion, philosophy, or science.

The great artist invents or discovers a perspective, a point of view, a style which opens a new and satisfying way of understanding an aspect of human experience. Subsequent artists develop, embellish, and alter this insight until some ultimately turn it into a cliché. The theologian or the prophet sees the world in a new way and followers develop it and transform it into dogma. The philosopher shows something new about the nature of things and it becomes common sense. The scientific innovator looks at a problem through new spectacles and eventually his work becomes a definition.

What I have said in the preceding paragraph is oversimple. Note that I might have paired "cliché," "dogma," "common sense," or "definition" with any of the list—art, religion, philosophy, science—and still have made sense. Likewise, politics insofar as it is a species of human artifice proceeds from—what shall we say?—"insight" to cliché, dogma, common sense, or definition.

Note carefully that I have suggested that politics—and art, religion, science, and philosophy—should be understood as "species of human artifice" and that each "proceeds" from something to something else. The notions contained in both these phrases are fundamental to all that is set down in these pages and I would not like to have them slip by as mere metaphor.

Man is a part of nature. He acts in his particular environment over time. Thus, human activity must be understood in terms of what we know about nature and natural processes and, above all, it must be understood as taking place in time. In an important way asserting that man is a part of nature and that human activity takes place over time comes to the same thing, because as natural science progresses it becomes increasingly clear that a fundamental feature of nature is its time dimension.

Let me now assert what I hope later to demonstrate, namely, that any attempt to understand politics by removing its time dimension creates distortion, often distortion so severe that the attempt ends in failure. The peculiarities of the past hundred years of the history of ideas have led a great many people to accept a dichotomy between history and science. The historian, it is said, tells the story of discrete events, while the scientist seeks patterns and generalizations which apply to any time and place. While there is considerable plausibility to this statement as a description of the behavior of historians and scientists, I do not accept it as it is often intended, that is, as prescriptive of what should be done. Thus, in arguing for the inclusion of the time dimension in political understanding, I am not defending traditional narrative history against social science. Nor am I defending the history of political ideas as ordinarily understood against the application of what is understood as scientific method to political understanding. What I do suggest is that any attempt to get at politics by a conceptual apparatus that excludes time misses the elemental feature of political activity.

The dichotomy between science and history is a trick played on us by the clichés, dogmas, definitions, and common sense of our immediate intellectual ancestors who in seeking insight into particular problems left us, often through no fault of their own, caught in a hardened residue from which we must struggle to escape.

These pages should be understood as the repetitive blows of a pickax against the wall of rigid ideas which keeps us from political understanding. The part of this wall upon which all else rests consists of ideas which are not distinctly political ideas at all; but because they are fundamental, we must start with them if later political observations are to make sense. Let us proceed then with a careful look at science, nature, time, and man.

PART ONE

The Universal Generalization Paradigm: a Critique

1
A Matter
of Perspective

The question of perspective is all-important. One must be clear from the beginning that "the facts" about politics or about anything else do not "speak for themselves." The point could be made by the use of any of a wide variety of examples, but it is perhaps sharpest in the context of scientific observation. Some years ago it was the fashion among theorists of perception to suppose that the human organism was bombarded by what were called "sense-data." Thus, what would in ordinary terms be called "looking at the sun" was described as receiving the impression "red patch now" or perhaps "hot red patch now" and then *interpreting* these raw data as the sun. Perception in these terms was a two-step process—receiving and interpreting.

Such an analysis, of course, tended to lend support to a radically positivist theory of knowledge. The data were simply there, simply given by nature and the test of truth or falsity was whether or not language described the data precisely. Contemporary men, armed with this understanding of the process of perception, were thus thought to be able, so to speak, to look the facts straight in the eye and to discard all of the metaphysical speculations of past centuries. There was, of course, a certain cost involved. A good many questions thought fundamental and important in the past were beyond the bounds of knowledge. Their importance was dismissed on the grounds that the questions were mere pseudoquestions, produced by distorted perceptions and distorted notions of the character of perception.

The assertion that man is capable of receiving raw sense data underlies the notion that "scientific knowledge" is a special kind of new knowledge, a sort of knowledge which is uniquely objective. Where the vision of previous inquirers was distorted by the presence of philosophical or religious presuppositions, the scientific observer was trained to look only at the facts themselves. Contemporary man, trained in the techniques of scientific observation, was said to be capable of what no earlier man could do, that is, to discover the simple truth about things.

Recent philosophical and historical research into the nature of scientific activity has, however, cast the greatest doubt on such proclamations of radical objectivity.* Hanson, for example, suggests that nothing can be just seen in the sense described by sense-datum theorists. Seeing always takes place in a context of perceptual underpinnings. What kind of sense, Hanson asks, does it make to assert that a geocentric astronomer and a heliocentric astronomer see the same thing when watching the sun rise? Imagine a volcanologist standing side-by-side with a primitive Polynesian watching the eruption of a volcano. The volcanologist *sees* a geological phenomenon of a certain intelligible sort. The Polynesian *sees* the wrath of a god. Surely it is a distinction foreign to the perceptual life of either to assert that they both "see" the same thing but "interpret" it differently. If you find this difficult to accept, let me invite you to try it yourself. Look at this simple drawing.

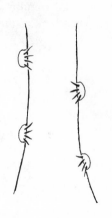

When I tell you that it is a bear climbing up the other side of a tree, do you engage in any interpretation or do you just *see* it that way? And, significantly, you see it that way irrevocably.

---

* See especially Norwood Russell Hanson, *Patterns of Discovery* (London: Cambridge University Press, 1959) and *The Concept of the Positron* (London: Cambridge University Press, 1963); Stephen E. Toulmin, *Foresight and Understanding: An Enquiry into the Aims of Science* (Bloomington: Indiana University Press, 1961); and Thomas S. Kuhn, *The Structure of Scientific Revolutions* (Chicago: University of Chicago Press, 1962).

Consider this rather more complicated drawing:

Now look carefully at the duck in the center who has only his head showing. Then look at the drawing below:

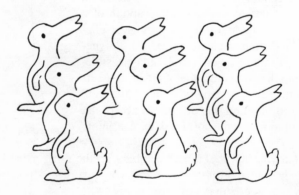

The "rabbit" in the center of this drawing is precisely congruent with the "duck" in the center of the previous one. Did you interpret the figure in the first drawing as a duck or did you just *see* it as a duck?

The final illustration which I would like to present is a more difficult one, but its impact is correspondingly more dramatic and forceful.

Here I must supply you with a verbal context. What looks at first glance like something taken from a Rorschach test is to be seen as a medieval representation of the bearded Christ looking directly at you. The upper border of the picture cuts off the top of the head at the eye brows. The lower border cuts across the chest.

This one, as I suggested, is rather difficult to see. Some people see it rather quickly; for some, hours are required; and some (as far as I know, a distinct minority) can never see it at all. The explanatory paragraph above sets the context and helps the seeing along, but as you will discover—if you haven't already—the *seeing* itself is instantaneous. What is most significant is that *interpreting* the patches of black and white (as some are inclined to do with an abstract sculpture or a contemporary poem) will not help at all. It is a matter of just seeing and not of interpreting.*

At the level of immediate, direct observation then, there is no such thing as seeing "reality" directly without help from a contextual underpinning—or as some psychologists would have it a *Gestalt*. The purpose of these pages is to achieve a clearer understanding of contemporary politics and insofar as understanding involves observing political events, perspective—as was noted earlier—is all important.

Significant as questions about the nature of observation may be, there is also a good deal more that must be made clear. The process of understanding events and circumstances necessarily involves a method or a

* The illustrations and the commentary upon them are adapted from Hanson, *Patterns of Discovery*, pp. 12–14.

style of understanding, whether that method or style be called scientific, philosophical, or whatever.

The problem that we have set for ourselves here is one of at least partially understanding what we have come to call the political activity of man. If (as has been suggested by examining the nature of immediate perception) perspective, "the way you look at a thing," is fundamental to understanding what a thing is or the way in which it works, then raising the question of the method or style of political understanding presents an extraordinarily important question—the answer to which is rather difficult to make clear.

The method or style of understanding that I propose to employ may properly be called "scientific." Man, as was asserted earlier, is a part of nature and he must be understood as such. To proceed in a properly scientific sense toward the understanding of an aspect of man's activity— in this case political activity—necessitates seeing man and his politics in the context of what Teilhard de Chardin has called "the whole phenomenon of man." Looking at the whole phenomenon of man demands, among other things, that in treating human behavior as an object of scientific inquiry we must not lose sight of the fact that part of human behavior has been the *creation* of scientific, artistic, philosophical, and religious activity.

The kind of thinking that presently advertises itself as the "scientific" understanding of human affairs is characterized by the borrowing of an analogy or a technique from a particular aspect of physics or biology and applying it to an aspect of human activity. It is thus called "scientific" in large measure—I would not argue that this is all there is to it—by virtue of the fact that it looks or sounds like something which is generally understood as scientific.

But it will not do to *begin* by abstracting out aspects of human interaction that look like, say, the interaction of cells and organs in a complete organism. It will not do just because perspective *is* all important and the perspective carried in with these analogies and techniques is false with respect to the whole phenomenon of man. (I would, however, be inclined to argue that certain analogies and techniques borrowed from natural science can be quite useful once a broader perspective is established.)

It is the question of where one *begins* that is important. Understanding the activities of groups of men as if they were functions of the parts of a machine or the parts of an organism makes it difficult, if not impossible, to properly account for those aspects of human activity that are unique to human activity.* Practitioners of this sort of enterprise usually recognize,

---

* Cf. Clifford Geertz, "Ideology as a Cultural System," in David Apter, ed., *Ideology and Discontent* (New York: The Free Press of Glencoe, 1964), pp. 47–76.

or more accurately perhaps, "feel" this difficulty and attempt to cover it by a variety of qualifying remarks.

Beginning with a perspective on the whole phenomenon of man and attempting a scientific understanding of an aspect of human activity means —paradoxically perhaps—that we must see man as that aspect of nature which created scientific understanding, something, of course, which no machine part ever managed. We must therefore look carefully into the nature of scientific understanding not just for one purpose, but for two. We must, first of all, try to discover a method or a style for the present inquiry, and secondly, by understanding scientific understanding we shall also come to discover something very important about the *object* of our inquiry, namely, man himself.

2
Science as
Human Artifice

British philosopher and historian of science Stephen Toulmin titled his 1960 Mahlon Powell lectures at Indiana University *Foresight and Understanding: An Enquiry into the Aims of Science.** I propose here to follow his general argument, inquiring similarly into the aims of science, with an eye to the answers to our two questions regarding the nature of science and the nature of the human activity which created science.

"Definitions," Toulmin suggests toward the beginning of his discussion, "are like belts. The shorter they are, the more elastic they need to be." There is a strong inclination most notable perhaps among writers of introductory science textbooks, philosophers attempting to give an account of science, and social inquirers trying to be "scientific" to present a short, concise definition of science.** Much can be learned, Toulmin suggests, by examining the problems created by these attempts at concise definition.

* Published under this title in 1961 by the Indiana University Press and later issued in paper cover by Harper Torchbooks.

** Thomas S. Kuhn to whose work we shall later be referring in greater detail

Suppose that one begins by attempting to characterize science simply and briefly in terms of a single aim. "The purpose of science," we might say, "is to discover reality," or to choose an example very often used by social science, "The purpose of science is to predict correctly." Either of these, or both of them together, make the question easy and if they are used as the theme of an introductory lecture in, say, a course in political science allow the instructor to get on without delay to the "substantive" aspects of the subject in question.

A great many natural scientists are inclined to leave such questions of definition to the philosopher in order to proceed with what are regarded as the important substantive questions of physics, chemistry, or biology. This attitude is on the whole hard to argue with. Social scientists, however, are under a particularly strong obligation to look critically at their own definitions of science just because they are borrowers and imitators. They inherit a tradition of social history or social philosophy which they are avowedly trying to make more scientific. Thus, what is understood by the term "scientific" becomes crucially significant.

Anyone who has been exposed to contemporary social science knows that the notion that "the aim of science is to predict correctly" is very widely accepted. Toulmin, very appropriately for our purposes, presents an acute critique of this definition. Before describing his argument, some prefatory observations about the notion of prediction as the aim of science are in order. Everyone, including I think the most enthusiastic adherents of this definition, would admit if pushed that correct prediction in the sense of simply accurately foretelling the future is not necessarily a mark of scientific activity. A fortune teller even if he turns out to be correct, would not be accepted by social scientists as one of their own. The argument, which I have heard more than once, that social scientists apply cautious "scientific" procedures because they are not men of genius capable of great insights might seem to belie this statement of exclusion, but it is nonetheless for the most part accurate.

---

opens his *The Structure of Scientific Revolutions* (Chicago: University of Chicago Press, 1962) in the following way: "History, if viewed as a repository for more than anecdote or chronology, could produce a decisive transformation in the image of science by which we are now possessed. That image has previously been drawn, even by scientists themselves, mainly from the study of finished scientific achievements as these are recorded in the classics and, more recently, in the textbooks from which each new scientific generation learns to practice its trade. Inevitably, however, the aim of such books is persuasive and pedagogic; a concept of science drawn from them is no more likely to fit the enterprise that produced them than an image of national culture drawn from a tourist brochure or a language text. This essay attempts to show that we have been misled by them in fundamental ways. Its aim is a sketch of the quite different concept of science that can emerge from the historical record of the research activity itself." Page 1.

Why, then, is correct prediction so closely associated with science in the minds of social scientists? The most important reason, it seems to me, is because social scientists tend to accept the style and procedure of classical Newtonian physics as a part of the very definition of science. The ideal of science epitomized by Newton consists in the discovery of general laws valid everywhere for all time in terms of which particular instances can be interpreted. Generalizations recognized as having something less than universal application tend to be understood as steps on the way to absolutely general laws. Prediction, it is argued, becomes crucial in this scheme because it is very often the case that only through a series of correct predictions can the generalization be validated. Thus, one "knows reality" in the sense that *x* and *y* are related in such-and-such a fashion only because *x* appears in such-and-such circumstances. Combine this with the powerful desire of social scientists "to know what is going to happen" and to be able to change things for the better and "correct prediction" emerges as *the* aim of social science and the definition of science in general.

All of this is important because social scientists have made "being scientific" a goal equal to that of describing the family or the political system or whatever and therefore, what "scientific" *means* will affect what we are told about the family or the political system. Again—although in a somewhat different sense than we used the phrase earlier—perspective is all important. With these observations in mind let us now, with Toulmin, have a good look at the notion of prediction.

Asking for the single purpose or aim of science, Toulmin suggests, is rather like asking for the purpose or aim of sport. One must immediately ask, "Does sport have a single purpose at all?" We are here back again to the problems of essential meaning with which we opened this discussion. Suppose it is suggested that, "The purpose of sport—of a game—is to defeat one's opponent." There are, of course, games, like solitaire, in which one has no opponent so that it hardly makes sense to talk about defeating him. We are faced with an unacceptable choice. We can define sport arbitrarily—"solitaire is not a game." But what reason could there be for excluding solitaire except the external one of purifying the definition? The other alternative is to qualify the original intention—"The purpose of *competitive* sport is to defeat one's opponent." This amounts to asserting that the purpose of competitive sport is competition, all of which advances understanding very little.

Toulmin contends that:

> By the time we can give a satisfactory characterization of the nature and aims of sporting activities, we shall be forced to abandon the original idea that sports have a *single* aim and purpose. So long as we rely on a portman-

teau answer this idea may appear innocent enough; yet if we want an answer with real substance its implications are unacceptable. Competitive games are of course competitive, and this fact determines the general *sorts* of purposes they will have. Yet in playing any particular game a man will be trying to do a large number of different things—to ace his opponent, tire him out, get him out of position, and so on—pursuing a range of different aims, any of which may contribute to his overall success. It will be his business to add to his score of points or goals in whatever way will best promote the general strategy of his activity. And we shall understand the game in question only when we understand in outline at any rate, this whole plural range of aims and purposes which a participant in it has to pursue.*

And science, suggests Toulmin, is like sport.

From the point of view of social science there is considerable irony in the directive which Toulmin here offers to the philosopher of science. He is objecting to the attempts of earlier philosophers to specify the nature of science in terms of one or a few abstract principles—"predictive success," "discovering reality," and so forth—and directing them instead to look in detail at what scientists actually do. Political scientists, objecting to the traditional abstract principles of political philosophy and constitutional or legal prescription, have accepted—in the process of trying to make themselves scientific—the abstract principles of the philosophers of science and used them as justification for looking into the actual real-life behavior of men in society. The irony finds its finest edge in the person of the scientific political scientist who dismisses with the utmost disdain accounts of a polity in terms of the abstract principles of political philosophy or constitutional prescription, but who himself justifies his science with the abstractions of a simplified philosophy of science.**

For alas, scientists too "behave," and science is a slice of life and not a machine which grinds out answers, equally valid, to questions no matter what the subject matter. It is worthwhile asking how does one judge a scientific theory? What makes a particular theory or hypothesis better or worse than another one? Again, the temptation is strong to answer with a brief and simple phrase: "The better theory is the one that explains more." Such an answer, however, puts us back just where we were before. Some of the activities of scientists, for example taxonomical or classificatory ones, do not even try to explain. A taxonomy can be right or wrong, helpful or misleading, and undeniably significant, perhaps even crucially significant, for a particular science while being related only by the slimmest threads to anything explanatory. But surely, *explanatory*

* From *Foresight and Understanding* by Stephen Toulmin. Copyright © 1961 by Stephen Toulmin. Reprinted by permission of Indiana University Press.
** Note again the remarks of Kuhn quoted on p. 13 above. His whole book *The Structure of Scientific Revolutions* supports this point.

theories are to be judged in terms of how well they explain. Competitive games are competitive and explanatory theories are explanatory.

We are driven a little further toward the conclusion that understanding science involves understanding in detail the ways in which these animals called men explain things and what explaining means to them. But, we might say, isn't this just where the notion of prediction comes in? This is what men are up to when they do science, getting correct predictions. Won't this kind of account allow us to sum this matter up? Let us have Toulmin himself speak to this point.

. . . Before we launch ourselves irrevocably into an empirical study of the methods of the different natural sciences, we must face one last question. 'Is there not *one* thing common to the different explanatory sciences, which unites all sciences more closely than all games? Are not explanations essentially means of making *predictions?* One need not get bogged down in tautologies, for a perfectly good and clear account can be given of what explanation entails. The purpose of an explanatory science is to explain— that is, to lead to predictions; and the merits of a scientific theory are in proportion to the correct predictions which it implies.'

This view of science has, in recent years, been very popular among philosophers, and it would be silly to ignore it. (If I now criticize it vehemently, that is partly because I once held the view myself.) In scrutinizing it, we must bear one question in mind. Replacing the terms "explanation" and "explanatory power" by "prediction" and "predictive success" will help us, only if doing so takes us on to a new level—one at which we begin to recognize, in real-life terms, the force and structure of explanation and explanatory power. There is no merit in trading one word for another equivalent one. We want to grasp the substance behind the words.

To anticipate: the predictivist account of explanation fails to help us in this. At first the statement "A successful explanation is one that yields many predictions" appears genuinely informative and illuminating. But it proves to be so only so long as we leave certain ambiguities in the key term unresolved. For the word "prediction" is in fact a very slippery one. It slides between two extreme uses: one naïve, the other sophisticated. In its most obvious and appealing sense, explaining and predicting are emphatically *not* all-of-a-piece; but, by hedging the term around with sufficient qualifications, we *can* at last use it to provide a definition of explanation. Unfortunately, the effect of all the hedging and qualification is to leave our original problems entirely unsolved. There proves, in the end, to be no substitute for a direct and detailed enquiry into the nature of explanation itself. And how we should embark on this enquiry begins to become clear in the light of the very contrasts between scientific "explanation" and simple, naive "prediction." *

* *Foresight and Understanding,* pp. 22–23. (Toulmin's italics.)

Prediction is indeed a slippery term and we shall have to be quite clear about what we mean by it. If we take it first of all in its literal sense, it means a "saying before," a prediction. This would seem to provide a good, clear quantitative test for a theory. The more correct pretellings, the better the theory. Such a test, however, rather clearly implies that a theory which yields no correct predictions is of no value. A good many perfectly respectable theories, most notably that of Darwin, have produced no straightforward predictions at all. No one has ever used the theory of natural selection to predict the appearance of a novel species, let alone verified his forecast. Yet to deny the explanatory power of Darwin's account would be absurd.

But, it might be objected, you are being too literal and too rigorous. It is really not necessary that a theory be able to predict all of the things that it can explain. Putting Darwin's theory to the test *has* involved prediction, but on what might be described as a smaller scale. When the Australians infected rabbits as a way of controlling the rabbit population, it was accurately predicted in accordance with Darwinian principles that a new strain of rabbits resistant to the disease would develop. So, it seems that in the final analysis the objective was prediction after all.

Do we, however, really want to accept the contention that after nearly a century in which Darwinian ideas were accepted because of their purely explanatory power, when someone makes a small-scale prediction—which might very well have been made without benefit of the whole Darwinian apparatus—that such a correct prediction is the high point of Darwin's science? Surely, one would have to have an extraordinarily powerful vested interest in the notion of prediction to do so. Vested interests in ideas can, of course, be nearly as strong as those in property or in status.

Lest we be accused of stacking the argument, we must recognize that the defense of prediction against an example like the Darwinian one has been put in a more sophisticated way. "Why should we," it is argued, "restrict the notion of prediction to its ordinary, vernacular signification. Why must we limit it to the future, to things that have not yet happened?" One can, it has been suggested, predict things that have already happened in the sense of anticipating what has not yet been discovered about the past. Palaeontologists, for example, have confirmed the usefulness of Darwin's theory over and over again by looking for and finding pieces of evidence which would not otherwise have been expected.

The force and the truth of this argument cannot be denied. Darwin's theory like several others of similar type *has* led investigators to past events which would not have been suspected on other grounds. And it is reasonable to call this prediction—reasonable in the sense that it does not torture the English language beyond recognition. However, it is perfectly clear that in so employing the term "prediction" the argument has shifted ground. We are no longer talking about prediction in the ordinary sense of foretelling the future, but rather about "inference" to any event whether past, present, or future. It is worthwhile pausing for a moment and asking yourself whether this argument, as revised, really supports what is intended when someone says "the test of a theory is whether or not it produces correct predictions."

Before reaching a firm conclusion with respect to the foregoing question, we should perhaps push a little further into the realities of scientific activity. Suppose that we ask what are the most striking examples of correct and precise prediction, in the revised sense, to be found in science? It would be difficult, I should think, to deny the title to the arithmetical techniques used to predict the motions of heavenly bodies and the times and heights of tides. This is accuracy and precision at its sharpest. And one can with equal ease calculate the position of Mars at the time of the birth of Christ and in the year 2000.

What is interesting is that some of the most successful techniques used for this purpose were developed through trial-and-error without a theoretical or explanatory basis at all. Yet, on the other hand, some quite respectable explanatory theory about this same subject matter produced little or nothing in the way of accurate prediction.*

This contrast is perhaps most clear at the time of astronomy's beginnings. Astronomy apparently had two independent origins. Five hundred years or so before the birth of Christ both the Babylonians and the Ionian Greeks were approaching the heavens in new ways which were destined to serve as foundation stones for the science of astronomy. Their respective approaches were, however, radically different. The Babylonians were extremely skillful in calculating the times and dates of astronomical events. The Greeks knew little of this until the conquest of Babylon by Alexander the Great centuries later. The Babylonians likewise exhibited a much more exact command of the calendar, and their techniques for forecasting the new moon and lunar eclipses were far beyond that of the Greeks of a comparable period.

Yet for all their predictive skill the Babylonians called the planets by divine names and had (so far as can be determined) no very significant ideas about the physical nature of the planets. By keeping detailed records of astronomical observations they were able to calculate the celestial motions by purely arithmetical means. They were able to extend this technique successfully to the movements of all the major planets and (as if to demonstrate for us their lack of theoretical sophistication) they tried less successfully to apply it also to earthquakes, omens, and plagues of locusts.

* Cf. Stephen Toulmin and June Goodfield, *The Fabric of the Heavens* (London: Hutchinson & Co., 1961), pp. 15–89. (Available in the United States in Harper Torchbooks).

The Ionian Greeks, on the other hand, were speculators. They conceived of circular tubes of fire with small holes through which the "stars" were visible. Black, nonluminous bodies which at various times obscured the light of the moon and sun were envisioned. They even speculated, interestingly enough, that the moon borrowed light from the sun and possessed no luminous powers of its own. As predictors, however, they were woefully weak.

In contemporary science we would expect virtues of both kinds, and eventually of course the Babylonian and Ionian strains in astronomy united. But at this early time the Babylonians, while they were skillful forecasters, lacked explanatory power; and the Ionians who were not, it seems, much interested in prediction were very much interested in understanding and explanation. If prediction is really as is suggested by the phrase "the test of a theory is whether or not it produces correct predictions," then surely we are driven in the direction of granting the title "scientist" to the Babylonians, who apparently saw no difference between the coming of the new moon and the coming of a plague of locusts, and denying it to the Ionians, who by all accounts invented theorizing, just because they couldn't predict.

Before you settle upon this as a satisfactory answer, note the fact that tidal and astronomical calculations are still carried on pretty much as they were by the Babylonians. Newton allowed us to *understand why* these techniques work, but his theories did not in themselves lead to more accurate prediction. Thus, if one is inclined to dismiss Thales, he may discover himself dismissing Newton as well.

It must be clear by this time that, with Toulmin, I am inclined to have strong reservations about the predictivist thesis. However, the discussion is not yet over, and we must allow the predictivist to make his final argument. We have up to this point treated prediction as always involving a dated, categorical assertion about the occurrence of a particular sort of event (whether, given the earlier modification, it is in the future, past, or present). But surely, it will be argued, this is too restrictive, because scientists can and do make hypothetical, conditional predictions as well. Some predictions, like those of eclipses of the sun or the moon, *are* categorical, but many others—indeed the most important ones—are of an "if *x* obtains, then *y* will follow" variety. Scientific experiments, after all, test "if-then" situations.

Again, this is an argument which must be granted, but notice how far we now are from the original proposition that "the test of theory is whether or not it produces correct predictions." We began by envisioning an external, independent test of a putative explanation, namely, the correct predictions it produces. "Here is the explanation—now let us put it to the test and see if it can predict anything or not." But the so-called hypothetical or conditional prediction just introduced is quite a different thing.

It is possible to call the "then *y*" in an "if *x,* then *y*" statement a prediction, but it is no longer an external test of the explanation but part of it.

We remarked earlier that Newton's theories allowed us to understand why the calculations concerning tidal changes and celestial motions were accurate. What does it mean to say that we understand why the computational techniques work, even though the Babylonians did not? Toulmin speaks to this point in the following way:

> We say this, mainly because we now have a number of general notions and principles which *make sense* of the observed regularities, and in terms of which they all hang together. Think how different Kepler's laws of planetary motion appeared after Newton. Kepler discovered that the orbit of Mars was, as near as he could tell, elliptical. He had some ideas of his own about the forces responsible for the planetary motions, but these provided no compelling reason why the orbit should have just the shape it did. So, the elliptical shape of the orbit was, for him, just a tiresome, obstinate, and arbitrary fact—"one more cartload of dung," which had to be brought into his system, "as the price for ridding it of a vaster amount of dung." Newton, by contrast, gave us a whole new set of conceptions, in terms of which Kepler's regularities ceased to appear as arbitrary facts. In the new theory, they all made sense and hung together, granted only a few plausible suppositions. Kepler told us: such-and-such *is in fact* the case. Newton showed us that, if we only supposed so-and-so, then on his principles Kepler's facts must be as they are ("Freely moving satellites, acted on by a single central inverse-square force only, *must* move in conic sections"). True: if Newton's initial suppositions had been unsound—if, for example, the space between the Sun and planets was not effectively empty—then his explanations would have been quite irrelevant. We should have had to find some other reason for the motions being those Kepler discovered. Actually, his hypotheses were perfectly plausible and made sense, not only of Kepler's laws, but of a great many other things as well. Yet notice: their merits were explanatory rather than predictive. They showed us what must happen *if* certain conditions were fulfilled, not what must happen *unqualifiedly.* They thus drew attention to an intelligible pattern of relationships between apparently unrelated types of happening— the ebb-and-flow of tides, the appearance of comets, the fall of stones, and the motion of planets. This "nexus" of regularities and connections was the thing that mattered most for Newton: it determined in his eyes, both what did happen in fact and what would have happened if conditions had been otherwise than they were. It formed a network of natural necessities, holding equally for the actual, and for unfulfilled conditions.*

In order to save the predictivist thesis, we can, as suggested earlier, argue that the Newtonian theory *predicts* the way in which the planets will move, given the conditional premises of an empty interplanetary space

---

* *Foresight and Understanding,* pp. 33–34.

and an inverse-square gravitational force. But in so doing the "prediction" becomes not the "external test of the validity of an explanation" that we began with, but merely another way of describing Newton's explanation. After looking at "prediction" in its straightforward sense of foretelling the future and after stretching and modifying the definition—like an elastic belt made to fit anyone—we are driven to two conclusions. In Toulmin's words:

> On the one hand, you can take the term "prediction" to mean the same as "explanatory inference"—but then the doctrine that the function of explanatory theory is to yield predictions leads straight back to the original, unhelpful tautology: "The purpose of explanatory theories is to explain." On the other hand, you can take the term to mean, simply, "forecast"—but then it turns out that the predictive success of a theory is only one test of its explanatory power and neither a necessary nor a sufficient one.*

All of this examination of prediction—necessary for our discussion though it is—does not, as far as it goes, really tell us what science is all about. Our earlier questions about the nature of science and of explanatory inference still stand. It would, however, be foolish not to recognize the ways in which forecasting is significant for scientific understanding. Forecasting, we may fairly say, is a craft, a skill, a technology which is related to science but is not its identifying feature. Forecasting, then, is like animal-breeding, smelting, or some aspects of medicine. While it is true —as we have noted—that a scientific explanation need not give rise to accurate forecasts or, for that matter, to any technological applications, it is also true that when a forecasting technique is successful, this in itself is a fact that science is called upon to explain. The history of science suggests that successful techniques—such as animal-breeding and smelting —were important stimuli for the development of scientific explanation. "It works, but why does it work?"

The calculations of the Babylonian astronomers are in fact a classic case. They, like craftsmen in the field of metalworking, operated on a purely empirical, trial-and-error basis and eventually they developed techniques that worked. On this basis we should not be surprised that they did not know the difference between the coming of the new moon and the coming of a plague of locusts. We can suppose that they judged their lack of success in the field of locusts in terms not unfamiliar to contemporary investigators of human affairs—"We don't have enough data!" Again, we may ask what do we mean when we say that Newton's theories made it possible to understand why the calculations of tides and celestial motions worked but the ones of earthquakes and plagues of locusts didn't?

---

* *Foresight and Understanding*, p. 35.

For the Egyptian astronomer, Claudius Ptolemy, predicting eclipses and the making of horoscopes were both empirical crafts and he argued that both were equally respectable scientific activities.

> Newton's theory explained why Ptolemy's eclipse predictions had been successful; but it gave us no reason for thinking that a man's personal fortunes could be forecast from astronomical signs. In this way Newton at last provided a substantial reason for distinguishing between astronomical and astrological forecasting. And how did he do this? By showing that the success of Ptolemy's astronomical constructions, too, tied in with his fundamental laws of motion and gravitation. *In terms of these ideals of natural order, facts which before appeared only arbitrary came to appear as natural and rational.**

This, I would argue, is the heart of the matter. It is "ideals of natural order" which are the kernel of science. Thus, the principal and controlling aims of science lie in the field of intellectual creation. Associated activities such as diagnosis, classification, and prediction are properly called scientific because of their connection with the ideals of order and explanation which make up the core of science.

We are now in a position to get some indication of sound answers to the questions with which we began this inquiry into the aims of science. It will be necessary to spend the next chapter pushing further into the character of these ideals of natural order, but we can at this point suggest the following: If our concern is to understand what we must do if this investigation is properly to be called "scientific," it is clear that we must pay attention to the sorts of ideals of natural order that we employ; and secondly, if we are to learn something about the creature who created science by examining the nature of the product, we must say that, again, ideals of natural order are the most important things created in the human development of science.

* *Foresight and Understanding,* pp. 37–38. (Italics added.)

3
Does Contemporary Political Science Rest upon a Mistake?

1

By way of introduction let me call your attention to the third volume, entitled *The Discovery of Time,* of a projected four volume series on *The Ancestry of Science* by Stephen Toulmin and his wife June Goodfield.* As the title suggests, this volume traces the development of the factor of time, as a significant explanatory device in the development of natural science. I cannot, of course, summarize the contents of the book in a few sentences. The point for political science can, however, be made sharply enough.

* *The Ancestry of Science* is published in the United States by Harper and Row, New York, New York. The first volume, on astronomy, called *The Fabric of the Heavens* appeared in 1961, and the second *The Architecture of Matter* in 1963. *The Discovery of Time* was published in 1965 and the final volume, *Science and Its Environment* is in preparation.

Toulmin and Goodfield show clearly that until the nineteenth century, with but a few minor and largely unheard voices in dissent, the search for scientific understanding was identical with the search for eternal verities (whether in the form of laws of nature, categories, axioms, or cause-and-effect relations) that lay behind the flux of experience. To put the same point another way, the search was for a set of variables which could be arranged in a static formula which in turn could explain the behavior of natural phenomena of a particular sort for all time. As a matter of the history of ideas one can, of course, locate the germ of this conception in Plato and find it further developed in Descartes, but the notion belongs above all to Sir Isaac Newton.

Newton is ordinarily described as a transitional figure. While it is true, the argument goes, that Newton used religious language and seems to reveal certain outmoded medieval sympathies, what is important is not these trivial peculiarities but his great breakthrough into intellectual modernity. His modernity, it is argued, lies in the fact that he explained in terms of natural and not supernatural causes. To repeat the language of the preceding paragraph, he found "a set of variables which could be arranged in a static formula (or formulae) which in turn could explain the behavior of natural phenomena of a particular sort for all time," but these were *natural* variables and required no notion of arbitrary, intervention by Divine Providence.

Developments in nineteenth and twentieth century astronomy, cosmology, physics, and chemistry together with, above all, the development of palaeontology, geology, and evolutionary biology have shaken the universal variable-universal generalization characterization of science almost to the ground. The principal factor which has brought about this change is, as Toulmin and Goodfield so aptly put it, the discovery of time. It is worth reflecting on the fact that Newton thought that the world was approximately 6,000 years old.

The overriding fact about modern and contemporary natural science is that it has become progressively more and more historical. Hear the testimony of physicist Carl Friedrich von Weizsäcker:

> But nature's appearance of being without history is an illusion. All depends on the time scale we use. To the mayfly whose life spans one day, man is without history; to man, the forest; to the forest, the stars; but to a being who has learned to contain within his mind the idea of eternity, even the stars are historic essences. A hundred years ago, none of us was alive. Twenty thousand years ago the forest did not stand, and our country was covered with ice. A billion years ago the limestone that I find in the ground today did not yet exist. Ten billion years ago, there was most likely neither sun nor earth nor any of the stars we know. There is a theorem of physics, the Second Law of thermodynamics, according to which events in nature are

fundamentally irreversible and incapable of repetition. This law I should like to call the law of the historic character of nature.*

While von Weizsäcker in 1949 can describe all of nature, including the account of it given by physics, as essentially historical, it is important to see that physics was last in developing this viewpoint. Toulmin and Goodfield, using in this context Platonic and Aristotelian thought as stylistic opposites, made the point in the following way:

> So the scientific priority and superiority sometimes claimed for physics— as laying the intellectual foundations, first for chemistry, and subsequently for physiology and even psychology—depends on our taking up an abstract, Platonizing point of view. When, by contrast, our search for understanding forces us to look at Nature in a more historical, Aristotelian light, the 'peck-order' is reversed: the science of palaeontology first achieved maturity, followed by geology, zoology and chemistry, and physics is still in its infancy.**

2

If we are to comprehend at the requisite level of sophistication just what is being said when it is asserted that "physics is still in its infancy," we must go into the question in some detail. I take it that the perspective of the utter vastness of time introduced into the modern understanding of the world in which we live by astronomy, cosmology, palaeontology, geology, and biology is well enough understood, at least in its general outlines. The implications for scientific understanding of the realization that the universe is some six billion years old as opposed to the biblically inspired notion that its age is a mere six thousand years are probably less well appreciated. There is no space to detail this matter here, but it is important to notice one quite general implication. Given the fact, obvious even in the seventeenth century, of the extraordinary complexity and diversity of nature, together with the assumption of a comparatively recent creation, a developmental understanding of nature was inconceivable. If there is order in nature—it was quite naturally assumed—that order must be in the form of a static pattern imposed on the stuff of things at the time of creation and remaining the same ever since. A cataclysmic intervention in ancient times like the Great Flood of the time of Noah could

---

* Carl Friedrich von Weizsäcker, *The History of Nature* (Chicago: University of Chicago Press, 1949), pp. 8–9. Copyright, 1949, by University of Chicago Press.
** From p. 271 in *The Discovery of Time* by Stephen Toulmin and June Goodfield. Copyright © 1965 by Stephen Toulmin and June Goodfield. By permission of Harper & Row, Publishers, Incorporated.

be admitted, but the idea—to which we have grown accustomed—of change so slow that it is invisible was literally inconceivable. Thus we have, in the context of natural science, inherited the phrase "law of nature" which was originally understood by Newton and others quite explicitly as a sort of legislative fiat issued by God at the time of creation.

Two points need to be clearly understood at this juncture if my eventual argument is to make sense. (1) The notion that all science is proceeding toward the goal of universal generalization epitomized by classical mechanics is quite wrong and understood to be so in 1969. (2) The universal generalization conception of science has been a long time dying— in some ways it is not dead yet—and it dies most slowly in the area where it has been most successful, that is, physics.

It would, it seems to me, be unreasonable to expect anyone to take my word for the correctness of these assertions, so let us allow Toulmin and Goodfield to speak for themselves at some length:

By the end of the nineteenth century this historical transformation had penetrated deeply into all fields of natural science—except one. In the geological, biological and human sciences * the *a priori* patterns of Greek philosophy had everywhere been displaced, but the new developmental categories influenced physics and chemistry much more slowly; and now, more than sixty years later, it is still open to question whether physics will ever become a completely historical science. The physical sciences had stood aside from the historical revolution which transformed the rest of natural science, taking it as axiomatic that certain aspects of the world remained fixed and permanent throughout all other natural changes; and though, by the mid-twentieth century, the list of these timeless entities—or "eternal principles"—as the Greeks had called them—is much shorter than it was in 1700, the existence of unchanging physical laws, at least, is still regarded as one enduring aspect of the physical world.

During the eighteenth century, the orthodox picture of physical Nature was stated by Isaac Newton at the end of the *Opticks*. This involved permanent features of five different kinds. First, there were the "solid, massy, hard, impenetrable, moveable Particles" that constituted the ultimate physical population of the natural world. It was inconceivable that these particles should change—at any rate until God resolved to terminate the present Order of Nature entirely. One of the crucial properties of these particles was their "force of inertia"—what we should now call their "mass"; and the second timeless feature of the world comprised those "passive Laws of Motion as naturally arise from that Force [of inertia]." By this Newton meant the laws

---

* The reference to "human sciences" at "the end of the nineteenth century" refers in the main to history and to the historical social science dominated by men like Comte and Marx. The nonhistorical social science with which we are now familiar is mainly a product of the twentieth century. Much more that bears on this point will be said later.

expressed in the axioms of his dynamics—and again, within the present dispensation, these laws of motion would not vary from age to age. Thirdly, Newton drew attention to "certain active Principles" associated with gravity, magnetism, electric attraction and "fermentation"—i.e., chemical reactions. These principles too, conformed to "general Laws of Nature" determined by God in the original Creation. For example, the "inverse-square law" of gravitation was part of God's timeless design for Nature, and neither the power nor the constant in the formula would vary without His direct intervention. The fourth permanent feature in the Order of Nature was the comparative stability of the planetary orbits; the fifth and last was the adaptive pattern in organic structure—what he called "the Uniformity in the Bodies of Animals." There was, Newton taught, no logical necessity about this particular picture. God was free to fix the laws of Nature just as He pleased, and it was quite conceivable that, in different parts of the cosmos, He had fashioned systems working according to other laws. But, so far as our own region was concerned, the Order of Nature had evidently been created on a fixed pattern of indivisible particles, laws of motion, active principles, planetary orbits, and organic uniformities.

Only in one respect was the Newtonian framework eroded during the eighteenth century. By the 1780s Kant's speculation that, on the astronomical scale, the creation of the natural order might still be in progress, seemed to be finding some real confirmation. William Herschel, who had built the most powerful telescopes yet available—with which he discovered the planet Uranus in the year 1781—soon satisfied himself that many of the so-called "fixed" stars were in fact in motion. Presumably gravitation acted not only within the Solar System, but between all members of the astronomical universe, and one result of this gravitational interaction could well be the concentration of stars into clusters and galaxies. With the help of a striking series of observations, he sketched out a possible account of the processes by which stars and star-clusters might evolve—the stars being formed by the agglomeration of faintly luminous inter-stellar dust and gas (which he first observed in 1790), then being drawn together into more and more compact associations, some of which eventually collapsed together in a final blaze of glory. This theory was immediately applied by Laplace to explain the origin and properties of the Solar System. That, too, could have formed by condensation out of an original diffuse cloud of matter, and its comparative stability required no other, supernatural origin. So, by 1800, one item had been removed from Newton's list of divinely-ordained principles and uniformities.

In other respects, however, the list remained unchanged. If anything, it became even more firmly established during the years that followed. For the "permanent Particles" of Newton's physical world were now identified with the "atoms" and "molecules" of nineteenth-century physics and it was widely accepted—at any rate among physical scientists—that God had originally created particles of some ninety different fixed kinds, which would persist unchanged throughout all the transformations of the physical world. In his address to the British Association in 1873, James Clerk Maxwell made the theological affiliations of this doctrine quite explicit:

No theory of evolution can be formed to account for the similarity of molecules, for evolution necessarily implies continuous change, and the molecule is incapable of growth or decay, of generation or destruction. None of the processes of Nature, since the time when Nature began, have produced the slightest difference in the properties of any molecule. We are therefore unable to ascribe either the existence of the molecules or the identity of their properties to the operation of any of the causes which we call natural. . . .

Natural causes, as we know, are at work, which tend to modify, if they do not at length destroy, all the arrangements and dimensions of the earth and the whole solar system. But though in the course of ages catastrophes have occurred and may yet occur in the heavens, though ancient systems may be dissolved and new systems evolved out of their ruins, the molecules out of which these systems are built—the foundation stones of the material universe—remain unbroken and unworn.

They continue to this day as they were created—perfect in number and measure and weight, and from the ineffaceable characters impressed on them we may learn that those aspirations after accuracy in measurement, truth in statement, and justice in action, which we reckon among our noblest attributes as men, are ours because they are essential constituents of the image of Him who in the beginning created, not only the heaven and the earth, but the materials of which heaven and earth consist.

The essential framework of the nineteenth-century physical universe thus consisted of fixed atoms, interacting by forces which conformed to fixed laws. Everything else was derivative and transient; and it is ironical that the effect of Dalton's and Maxwell's theories of matter was to freeze the central categories of physics and chemistry into a rigidly a-historical form at the very time that biology was being transformed into a mature, historical science by the publication of Darwin's *Origin of Species*.

During the twentieth century, the list of changeless physical entities has drastically shortened. The discovery of radioactivity raised the first doubts about the permanence of the atom, and this was only the first pebble initiating an intellectual avalanche (see Chapters 12 and 13 of *The Architecture of Matter*).* Though the physical world, as envisaged in modern quantum theory, is still composed of so-called "fundamental particles," these represent a much more transient and changeable population than either Newton or Maxwell could have contemplated. All of them, indeed, are transformed in appropriate circumstances into others, and by now there remain no "eternal units" of matter or energy at all. Some of the physicists' particles may seem more fundamental than others; but it begins to appear that their interrelations will be best understood when they are seen as by-products—temporary configurations in the fields of force which are defined by the mathematical equations of physical theory.

Newton's original five categories have thus been cut down to one: the

---

* The second volume in *The Ancestry of Science* series cited earlier in this chapter.

fixed laws of Nature. It is these whose permutations—in theory—govern the whole kaleidoscope of matter and energy which is the scientist's object of study. Given an appropriate and fixed system of equations, one can hope to explain physical processes from the sub-atomic up to the molecular, and macroscopic, and even the astronomical scale; and this can be done without exempting any physical objects whatever from the flux of time—which means accepting that there *are* no permanent material "atoms" in the original sense. For contemporary physicists, indeed, the particles making up ordinary matter are so many temporary consequences of the equations governing the fundamental force-fields, just as for Laplace the structure of the Solar System was a consequence of universal gravitation. The outstanding question now is, whether the laws of Nature themselves—the last a-historical feature of the physicists' world-picture—will in their turn prove to be subject to the flux of time.

The only region of science in which this question can seriously be raised at the present time is physical cosmology, for there we are concerned with processes taking place over thousands of millions of years, during which any slow alteration in the fundamental laws of Nature might become evident. In any case, the character of cosmological theory has at all stages been affected by changes in the central categories of contemporary physical theory; and the encroachment of evolutionary ideas into physical thought during the twentieth century shows up clearly in this, the most "historical" of the physical sciences. If we follow out the development of cosmological ideas over the last 150 years, we can watch this historical consciousness permeating the science, and examine the nature of the evidence which may eventually remove the last timeless features from our conception of the physical world.*

At this point Toulmin and Goodfield enter a rather extended discussion of stellar evolution which, while extremely interesting in itself, is not immediately relevant to our present concern. This discussion, so far as the matter of evidence is concerned, centers upon the red-shift in stellar spectra which has typically been interpreted as an instance of the so-called Doppler effect, that is, as evidence that the light source is moving away from the observer. The discussion is concluded as follows:

One further point must be mentioned in conclusion. We saw earlier in this chapter how the permanent framework of Newton's physics, with its five distinct categories of changeless entities, was progressively eroded, so that by the mid-twentieth century only the "laws of nature" defining the fundamental "force fields" still preserve their timeless character. But, looking at these present-day heirs of Newton's laws of motion and gravitation, one must ask one further question: Will these laws of nature remain for ever as the unchanging framework of the physical scientists' world-picture? This question arises both as a natural extension of the historical changes we have been

* Toulmin and Goodfield, *The Discovery of Time*, pp. 247–250.

studying, and because it has a direct connection with the present phase in cosmological thought.

To relate it back to our earlier discussion: the phrase "laws of nature" acquired a central position in European natural philosophy only in the seventeenth century. It entered the European tradition in a very particular historical context, and it had from the outset certain specific theological associations. The men who conceived the mathematical formulae concerned called them "laws" because they saw in them the "rules" by which the "behaviour" of material things was "governed." At this initial stage, the legal analogy implicit in the word "laws" was something more than a metaphor; and, since God Himself was regarded as changeless and eternal, it was presumed that the "laws of nature" which were an expression of His will were correspondingly fixed in their form. So the rational framework of the natural world which they called "the Order of Nature" was defined, and in justification it was sufficient for Newton that

> It became Him (God) who created all material Things to set them in order. And if He did so, it's unphilosophical to seek for any other Origin of the World.

The order of Nature was frozen by God's rational choice.

By now, however, this particular theological framework has lost its hold over us. Like so much else in the static world-picture of the seventeenth and eighteenth centuries, the eternal fixity of physical laws can scarcely be assumed any longer without examination. In any case, there are some scientists who have positive reasons for believing that this fixity, too, may be an illusion. Certainly, if the laws of nature were changing slowly enough, there would be no practical way of detecting this fact—even over millions of years—and in that case the question whether they were changing would be purely academic. But perhaps these changes in the physical processes occurring at different cosmic epochs are, after all, *not* undetectable; and perhaps we may even have already in our grasp—unrecognized and unappreciated—evidence to show their nature.

The red-shift in stellar spectra might itself be just that evidence, and is interpreted as such in one of the more unorthodox explanations offered for it. The more distant the galaxy from which a light beam comes, the more remote in time were the physical processes by which it was emitted. Now, if the numerical constants in the laws of nature governing these processes "drifted" slowly with the passage of time—e.g., if Planck's constant, h, were slowly increasing—this would produce the familiar red-shift, even though the distant galaxies were not receding at all. The apparent "recession of the galaxies" would be an illusion, produced by slow secular change in the laws of nature, and the "primaeval atom" would disappear from cosmological history.

This possibility is mentioned, not because the positive evidence in its favour is particularly strong, but simply to complete a balanced account of ideas about cosmic history. At a theoretical level, a Universe whose laws are changing from epoch to epoch is no longer an idle fantasy. The physicist P. A. M. Dirac, for example, has been exploring the intellectual consequences which follow when physical forces of different kinds (e.g., electrical and gravitational) vary slowly in their relative strengths: in this way, he links the atoms, with their electrical forces, to the galaxies, with their gravitational forces—the microcosm to the macrocosm—in a manner which changes through cosmic history.

We saw in *The Architecture of Matter* that the future development of subatomic physics is as open today as it has ever been. Nor can we be any more certain about the pattern of future cosmological theory. Though, during the last fifty years, our understanding of stellar evolution has been a great intellectual advance, which in some form or other will certainly survive in our world-picture, many profound and difficult questions are still not even half answered, and one of these is central: Do all the fundamental laws and constants of physical theory retain their forms and values eternally, without any modification? About this we can no longer be certain. We may yet be on the threshold of the greatest of all scientific revolutions.*

3

It is necessary, I would suggest, to think very hard about the force of the foregoing remarks by Toulmin and Goodfield. What is being asserted is that the nature of the real world has forced natural scientists, including the physicists, to give up a great many of their most cherished assumptions. More than this, it has forced a full-scale reexamination of what it means to be scientific. There are two points here. (1) Among scientists themselves what is involved in the understanding and the explanation of phenomena has changed. (2) Among those who think and write *about* science, most notably the philosophers and historians of science, a reevaluation of fact and interpretation has become a necessity. I should like to discuss both of these points with an eye to their relevance for political understanding.

The Newtonian image of the world (which persisted certainly to the end of the nineteenth century, as the earlier quotation from Clerk Maxwell clearly shows, and into the twentieth century as well) gave rise to what might be called the prediction-generalization account of scientific explanation. The prediction-generalization thesis suggests that if the relevant variables are isolated and that if they are theoretically linked in such a way that correct predictions result, the phenomenon in question is perforce understood and explained. What is perfectly clear when this notion is examined at all closely is that it presupposes the existence of a permanent constellation of variables which, when they are discovered, properly measured, and properly related yield correct predictions. The correct predictions prove that the variables have been properly isolated, measured,

---

* *The Discovery of Time,* pp. 263–265.

and related or, in other words, that the correct predictions establish the explanation.*

This is what Newton did with the variables mass and distance, and he thus explained at least in part an important aspect of celestial mechanics. But Newton, as we have already noted, had an extremely limited conception of time and he managed to pick out (and I suspect that this is no accident) a set of phenomena whose significant time dimensions were a great distance from any immediate application of his theory. Thus, given Newton's choice of phenomena he did not have to worry about time.

Newton's science and that of his many great successors, presupposing as it did a permanent set of variables, could be extremely successful while at the same time fitting—more accurately perhaps, *defining*—the prediction-generalization model. Success begets emulation, and the prediction-generalization account tended to become the very definition of science. The continuing emergence of data which pointed toward cosmological, terrestrial, and biological evolution over fantastic periods of time produced a group of scientists who, try as they might, simply could not actually *use* the prediction-generalization approach in their work even though their inclination to do so was strong.

This set the stage for what might be called a "paradigm change" or in more conventional language a scientific revolution. "Paradigm" as I want to use the term here is more or less synonymous with words and phrases like "ideal of natural order," "ultimate pattern of explanation," "explanatory world view." I would not want to say that it is equivalent to "*Weltanschauung*" or "metaphysic," but it does share in large measure the breadth of these terms. Similarly, it has a good deal in common with what Stephen Pepper calls "the root metaphor" in his book *World Hypotheses*.** Speaking functionally (and, I fear, somewhat vaguely) it is the paradigm that determines what counts as an explanation.

The meaning of "paradigm" in this context is perhaps best indicated by example. Let me again ask Toulmin for help, because his illustration seems to me better than anything I am able to dream up on my own.

Let us start with an elementary illustration. Suppose we set out to compare cooking with ripening. The question arises: is it more illuminating to com-

---

* I, of course, do not mean to suggest that correct predictions are the only test to which proposed explanations are put as a *practical* matter. Elegance, compatibility with other theories, intuitions concerning the adequacy of the explanation, and so on, are all a part of what the working scientist actually does. What I do contend, however, is that under this conception of scientific explanation correct prediction is the ultimate test of adequate explanation although, of course, as we noted earlier, in a good many cases the fact of correct prediction is not an altogether sufficient reason for regarding an explanation to be adequate.

** Stephen Pepper, *World Hypotheses* (Berkeley and Los Angeles: University of California Press, 1942).

pare the changes produced by cooking to the process of ripening, or to explain ripening in terms of the effects of cooking? Nowadays, we should probably plump for the latter: we should set about explaining the obvious, visible changes which take place when an unripe ear of wheat turns into a ripe one, by referring to invisibly small physical and chemical changes within the ear.

"Ripening?" (we might say) "Well, you know about cooking—how, under the action of heat in the oven, structural alterations are produced in the tissues of a steak—as a result of which it changes in colour and texture, and becomes easy on the jaw. Ripening is a comparable process. The heat of the Sun, like that of an oven, once again stimulates within the ear of wheat structural modifications whose consequences show up in its colour and texture. To begin with it was green and hard; but now, as a result of these structural changes, it becomes golden and softer."

The full story is, of course, much more complex than this, and probably has not yet been worked out. For the present purpose, however, the details are not important: it is rather the direction of thought—from physiology to physics and chemistry—with which we are now concerned.

In Aristotle the direction of thought is reversed. He does not set about explaining ripening by comparing it with cooking: rather, he works the other way round. Ripeness is all; material changes tend of themselves towards that goal; and artifical "concoction" can only speed up the normal, uninhibited processes of Nature.

"Cooking?" (he might say) "Well, you know about ripening—how, as the weeks pass, the seeds germinate, the infant seedlings turn into the adolescent stalks, and the plants finally come to their natural maturity— the innate qualities of the adult plant all the while coming to light, as the process of growth and maturation pursues its natural course. Cooking is a comparable process. The raw steak is no *Filet Mignon* to begin with; but it is capable of developing into one, if subjected to the appropriate environmental conditions; for, under these conditions, it is given the opportunity to manifest in fact all the inherent tenderness and succulence of which it is capable."

For Aristotle, cooking and ripening were both forms of "concoction," but ripening was the more typical and self-explanatory of the two. In each case the inherent qualities of an "immature" body were brought out, by subjecting it to the appropriate degree of heat for the appropriate length of time; so that cooking was, so to speak, a kind of artificial ripening. To understand a thing's material nature and constitution meant, in consequence, to recognize what it was capable of developing into, either naturally and of itself, or artificially if appropriately treated. Chemistry was thus subordinated to physiology, instead of the other way round.*

---

* Toulmin, *Foresight and Understanding* (Bloomington: Indiana University Press, 1962), pp. 67–69.

What is at issue here is an "ideal of natural order," a paradigm of explanation. If one understands it to be the natural order of things that they "grow" from potentiality to actuality—to use the Aristotelian terminology—then it is quite obvious that "cooking" should be explained in terms of "ripening." If you regard this example as farfetched, remember that for centuries men thought that metals grew "in the womb of the earth" from lower to higher states, and that the natural engineering consequence—that is, alchemy—of this understanding was a wide variety of attempts to accelerate this growth by the use of such methods as "seeding" molten lead with gold dust for the purpose of "breeding" more gold.

A paradigm is controlling in explanation in the sense that it defines what is "natural," what "stands to reason." Perhaps another example will help us to see clearly what is involved here. Consider the changes in the understanding of the motion of physical bodies epitomized by the theories of Aristotle, Galileo, and Newton.

Aristotle looked at the world as it was in his experience and noted that physical bodies do not move unless something moves them. He, thus, tried to state the relations between force applied, weight to be moved, and time necessary to do the moving. He thought in terms of real problems, like pulling a wagon through a muddy field. He, for example, concluded—quite correctly as far as he went—that the distance a body can be moved in a given time will be directly proportional to the effort employed and that the distance traveled in a given time will vary inversely with the amount of resistance (for example, in a muddy field) offered to the motion. He had no time for the altogether bizarre situation of a body moving without any resistance at all.

Galileo, however, for a variety of complex historical reasons was willing to consider the idea that rest and continuous motion were equally "natural" for physical bodies. Resistance, thus, became an external force with respect to the motion of bodies and not something which was always built into the situation, as it was for Aristotle. Armed with the notion of a body moving in a vacuum—a new "ideal of natural order"—Galileo went on to make what turned out to be significant advances in the field of physics. It is interesting to note, for example, that Galileo's famous experiments with the pendulum were made possible by virtue of the fact that he could regard the "slowing down" of the pendulum's swing as the consequence of an external—and thus theoretically irrelevant—force, namely the resistance of the air. Earlier experimenters, equipped with an Aristotelian paradigm, saw the pendulum as a body striving to come to rest near the center of the earth but impeded in its progress by the chain or rod upon which it was swinging. Thus, the question of the "period" of the pendulum, irrelevant and unseen for the Aristotelians, became a matter of great significance for Galileo.

But, of course, Galileo was still not Newton and did not come to Newton's substantially more majestic conclusions. There are no doubt a multitude of reasons for this, but there is one which is particularly pertinent to our present discussion. Galileo's idea of natural motion, while it allowed for continuous motion, was not rectilinear but circular. For, according to Toulmin:

"What he [Galileo] envisaged as his ideal case was a ship moving unflaggingly across the ocean along a Great Circle track, for lack of any external force to speed it up or slow it down. He saw that uniform motion could be quite as natural as rest; but this 'uniform motion' took place along a closed horizontal track circling the center of the Earth; and Galileo took such circular motion as entirely natural and self-explanatory. He does not seem to have regarded the ship as constrained by its own weight from flying off the Earth on a tangent—the image which can clearly be found in Newton." *

Newton changed his paradigm, his ideal of natural order so that it included natural motion *in a Euclidean straight line.* As a consequence the motion of the planets around the sun became something that had to be *explained,* rather than taken as natural. The results which followed from this new posing of the problem made Newton into one of the greatest figures in science.

4

Before moving to a direct discussion of certain features of contemporary political science, let us take the time to make two observations on our discussion so far. We have first of all noted that the nineteenth and twentieth centuries have produced a good deal of friction between the older, static "law of nature" understanding of scientific explanation and the new historical, developmental paradigm for scientific explanation. Explanation has become less and less a matter of "it obeys a general law" and more and more a matter of "it develops over time in such-and-such a way."

In the second place—and this can be equally instructive—the historians and philosophers of science, as we have seen by presenting their views, are changing their understanding not only of natural science itself but of their own enterprise as well. The conventional stance of the historian of science has been to regard "what we know now" as science and the history of science simply as the accumulation of this information. As Thomas S. Kuhn puts it:

* *Foresight and Understanding,* pp. 54–55.

If science is the constellation of facts, theories, and methods collected in current texts, then scientists are the men who, successfully or not, have striven to contribute one or another element to that particular constellation. Scientific development becomes the piecemeal process by which these items have been added, singly and in combination, to the ever growing stockpile that constitutes scientific technique and knowledge. And history of science becomes the discipline that chronicles both these successive increments and the obstacles that have inhibited their accumulation.*

The historian of science then has two tasks. First, he must show when and how the facts were discovered or the techniques invented, and secondly, "he must describe and explain the congeries of error, myth, and superstition that have inhibited the more rapid accumulation of the constituents of the modern science texts." **

This description, with certain important alterations, might very well be applied to a great many courses taught and books written in what is called the history of political thought. "Who had what insight when?" is the controlling question. Teachers of the history of political thought often use the answers to this sort of question as a way of showing that contemporary political scientists have discovered nothing new. On the other hand, many contemporary political scientists, having been instructed in the history of political thought by an "insight chronicler," tend to see the so-called insights as shallow and buried in a heap of "error, myth, and superstition."

Kuhn continues with a passage that can be instructive to both kinds of students of politics. "In recent years," he argues, "a few historians of science have been finding it more and more difficult to fulfill the functions that the concept of development-by-accumulation assigns to them. As chroniclers of an incremental process, they discover that additional research makes it harder, not easier, to answer questions like: When was oxygen discovered? Who first conceived of energy conservation? Increasingly, a few of them suspect that these are simply the wrong sorts of questions to ask. Perhaps science does not develop by the accumulation of individual discoveries and inventions." †

Thus, the historian of political thought can learn something. Who said it first is not necessarily what counts. Before the contemporary political scientist leaps from his chair in triumph, however, we must allow Kuhn to continue.

---

* Kuhn, *The Structure of Scientific Revolutions* (Chicago: University of Chicago Press, 1962), pp. 1–2.
** *The Structure of Scientific Revolutions*, p. 2.
† *The Structure of Scientific Revolutions*, p. 2.

Simultaneously, these same historians confront growing difficulties in distinguishing the "scientific" component of past observation and belief from what their predecessors had readily labeled "error" and "superstition." The more carefully they study, say, Aristotelian dynamics, phlogistic chemistry, or caloric thermodynamics, the more certain they feel that those once current views of nature were, as a whole, neither less scientific nor more the product of human idiosyncrasy than those current today. If these out-of-date beliefs are to be called myths, then myths can be produced by the same sorts of methods and held for the same sorts of reasons that now lead to scientific knowledge. If, on the other hand, they are to be called science, then science has included bodies of belief quite incompatible with the ones we hold today. Given these alternatives, the historian must choose the latter. Out-of-date theories are not in principle unscientific because they have been discarded.*

Once the philosophers and historians of science see the controlling character of the paradigm in scientific explanation they find it impossible to dismiss the theorizing of Aristotle or Newton just because their ideals of natural order are different from those held today. Such a dismissal becomes possible, as Kuhn accutely suggests, only when one assumes that past paradigms, because of their theological or philosophical associations, are "myth" and "superstition," and that on the other hand, contemporary scientists simply look the facts "straight in the eye." Accurate interpretation of past explanation is much more complicated than this, and neither ancient nor contemporary science nor the relationship between the two can be properly understood in terms of a dichotomy between "superstition" and "empiricism." The old saw which lies at the root of this dichotomy, that Copernicus and Galileo began modern science because they just "looked at the facts" while previous men were mired in the swamp of theological superstition, is plainly not true as writers like Kuhn, Toulmin, and Hanson have clearly shown.

Yet attempts have been made—many of them successful—to organize the study of political science around this very dichotomy. A series of books has lately been inaugurated with the pronouncement that after two thousand years of wandering in the wilderness an "empirical" political theory has finally been developed. In some major graduate schools the study of political theory has been divided into what is euphemistically called "traditional" political theory and "empirical" political theory. The separation is justified in the name of a philosophically and historically crude dichotomy between "fact" and "value," between "empiricism" and "superstition," between "science" and "philosophy."

Two very sensible and practical things can be done about this problem by professional students of politics. The history of political thought ought

* *The Structure of Scientific Revolutions,* pp. 2–3.

to be taught not just in terms of "who had what insight when" or "who favored what when" but also in terms of the development of political understanding, indicating and evaluating parallel developments in other fields and especially in the natural sciences. The work of Kuhn, Toulmin and Goodfield, Hanson, Hall, and others provides an excellent set of tools for this kind of teaching if someone would pick them up and use them. Students encountering contemporary political science might then be confronted not with an ideology to be accepted or rejected as they too often are now, but with an intellectual development that they could critically evaluate in its own terms.

Secondly, political investigators who want to make themselves more scientific must make a greater effort to discover what in fact being scientific means. Kuhn speaks of "the image of science by which we are now possessed" as having been drawn "mainly from the study of finished scientific achievements as these are recorded in the classics" and "the textbooks." ". . . A concept drawn from them," he continues, "is no more likely to fit the enterprise that produced them than an image of a national culture drawn from a tourist brochure or a language text." Hanson, Toulmin, and others speak to much the same point, but social scientists continue to talk textbook science and suppose it to be revolutionary. Much is made by political scientists of emulating the science of the sociologists and the psychologists, but a close look will reveal that they are on the whole beating precisely that dead horse of textbook science which Kuhn describes. What is important, as we have gone to some lengths to demonstrate, is not the techniques or the language but the paradigm of explanation, and it is the choice of paradigm that advances or retards understanding. With all this in mind let us now turn squarely to the question of *political* understanding.

4
Paradigm and
Political Understanding

1

If there is a single obvious fact about recent American political science, it is that a considerable portion of the discipline has undergone, and is still undergoing, a methodological revolution. No one who knows the field would deny that political science for the last thirty-five years or so has striven to become more scientific. It would also be agreed with a nearly equivalent measure of certainty that the "scientification" of political science has almost totally, as I suggested earlier, been a borrowing process.

The general theory of inquiry has surely come from the natural sciences. There has, however, been very little direct contact between physics, chemistry, and biology and political science. Scientific instruction has been at the hands of psychologists, sociologists, social psychologists, and in some measure economists and anthropologists. Insofar as political science has had contact with the natural sciences (and this is also in considerable measure true of other social sciences) it has been through the writings of historians of science and philosophers of science or natural scientists acting as historians or philosophers.

The fact that political scientists have not invented their own science but have become scientific through intermediaries is, we have argued, a matter of prime importance. As all political scientists know, greater attention to scientific method led to greater attention to how political actors actually behave and comparatively less to abstract ideals. What most political scientists, apparently, do not know is that the philosophy and history of science is presently under a comparable challenge, turning its attention away from abstract ideals of scientific method and toward a consideration of what scientists actually do.

Science as opposed to the natural world itself is a human institution. It compares in a great many significant ways to what we can call, for want of a more inclusive term, a polity. Viewed as itself a "thing" in the world, science is composed of equipment and documents both of extraordinary variety, and above all it is made up of men both numerous and extremely various presumably linked together by some common concern or concerns. Like a polity science has its power structures, opinion leaders, conservatives and liberals, revolutionaries, and, if you will allow the term to be stretched a little, even its ideologies.

No political scientist, and certainly no modern political scientist, would allow even a single polity like the United States of America to be summed up in a few abstract principles. Not a few of those same political scientists are quite willing to treat science in this way, even, it should be said, in graduate scope and methods courses where a few chapters from a standard text on scientific method is taken as sufficient answer to the question "What is science?" This presumes that it is possible to abstract a set of clear-cut, common principles called "scientific method" from the complex of activities labeled "science." It is just this presumption which contemporary history and philosophy of science calls into question.

2

In order to make a particular introductory point let us for a moment compare science to a sort of Neapolitan ice cream pie like the one pictured on the page opposite. If the solid section is chocolate, the center section vanilla, the cross-hatched section strawberry, and sections A to H bounded by dotted lines are possible pieces, it is clear that one's notion of the pie will depend very much on how he cuts into it. The example is a homely one, but the point needs to be made with a sledgehammer. *One's view of science depends very much on how and for what purposes he cuts into it.*

It is a commonplace that modern science was a revolutionary development in Western thought. It involved the rejection of old conceptions and

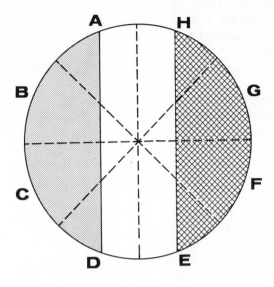

alternatives of inquiry in favor of new ones. The scientists themselves when acting as scientists rejected old explanatory models with respect to some aspect of the natural world in favor of new ones, but—writing in the late nineteenth and early twentieth centuries—philosophers, historians, and scientists acting as philosophers or historians stressed the mode of thought which differentiated the modern scientist from the natural philosopher of former times. The practical question for commentators on science was how to explain the great success of the new science. What was wrong with the old methods and what was right with the new? The intellectual question was what can be learned about proper methods of human thought by seeing what scientists have done.

It seemed clear that what differentiated modern scientists from previous investigators was their scrupulous attention to the facts. A man who got a piece of our Neapolitan pie which was eighty percent chocolate would be likely to think of the pie as primarily a chocolate pie. In the same way commentators coming to science with these questions in mind saw factuality as *the* characteristic of science. Speculation might play some role ("there is a little vanilla in the pie") but verification by testing against the facts is far and away the most important consideration.

The classic story which illustrates this perspective is the traditional account of the Copernican revolution. Astronomy prior to Copernicus, it is suggested, was bound up with a great many irrelevant metaphysical and theological considerations. For the early natural philosophers common sense dictated that the sun revolved around the earth, and religion held that this must be so because the earth and its most important inhabitant were God's highest creations and therefore occupied the center of things. Copernicus looked at the facts, ignoring religious teachings, and judged that this could not be so, causing, thus, no small amount of turmoil and beginning the emancipation of the mind of man.

This point of view, what might be called the "verification perspective" on science has for the past fifty years or so had wide circulation. Like every significant intellectual development it operates at a variety of levels. It is the common sense view of science; it obviously has the adherence of most social scientists and many natural scientists; it governed the writings of many early philosophers of science; and it had its most precise statement in the technical philosophizing of the early Wittgenstein and the Logical Positivists.* Wittgenstein in his *Tractatus Logico-Philosophicus,* written during World War I, inquired into the question of "how statements mean" and concluded that a statement has meaning if its structure supplies a picture of the corresponding state of the external world. Following Frege and in some measure Russell he admitted definitional structures like mathematics into the realm of meaningful discourse but, of course, denied them empirical meaning. These views were summarized and given a wide audience by the philosopher A. J. Ayer with the publication of *Language, Truth, and Logic* in 1936.

It is not unreasonable to see Ayer as having distilled the essence (to use language of which he would hardly approve) of the then current conception of scientific activity. The verification theory of meaning which is central to the doctrine of *Language, Truth and Logic* is the ultimate statement of what we earlier called the "verification perspective." The general tone of Ayer's exposition is most instructive for the point that we are trying to make here. The book is abrasive, highly critical, polemical —in a word, it is "negative," concerned largely with the debunking of prior philosophy, particularly metaphysical speculation which Ayer chooses to describe with no milder term than "nonsense."

Intellectual revolutions are very often initiated with largely negative statements. As Louis Hartz has observed, Karl Marx, the great theorist of socialism, spent ninety percent of his time talking about capitalism. Marx was long on devastating criticism of the old order but short on description of the new and better society. In much the same way the "verification perspective" is negative in tone; that is, by way of being in favor of hard facts it is involved mainly with debunking previous speculations.

---

* I have no doubt that Wittgenstein's views and intentions were distorted by the Positivists, but he nonetheless entered the marketplace of ideas with the Positivists. More later on Wittgenstein.

3

Let us now have a careful look at a description of the development of scientific psychology by a prominent psychologist, Charles E. Osgood.* We perhaps do not need to say that the psychologists have been among the principal science teachers for political scientists.

According to Osgood, behavior theory in the broadest sense is made up of hunches about how the nervous system operates to generate the lawful relations we observe among and between stimuli and responses. He states:

> Over the past century we can trace a gradual refinement in the rules of procedure whereby psychologists make and test these hunches, and this trend toward increased rigor in theorizing has been paralleled by similar development in the social sciences.
>
>   In nineteenth-century psychology the characteristic procedure in theorizing was to simply postulate a new entity or mechanism whenever some new regularity was discovered. Whenever something needed explaining a new explanatory device was stuck inside the little black box, and it rapidly became chock-full of ill-assorted and ill-digested demons. For every nameable phenomenon of human behavior a different "faculty" would be posited to explain it; for every nameable motive, a different "instinct" would be listed as its explanation. And, at least for communicating with patients, Freud had big, flat-footed Superegos stomping around on red, slippery Ids, while cleverly anxious little Egos tried to arbitrate. Thus, as suggested in the figure below, the little black box was filled with a wonderously diverse collection of explanatory devices, just about as many as there were things to be explained. This could fairly be called "junk-shop psychology." While it had the advantage of free exercise of often brilliant intuition, it had the disadvantage of complete lack of parsimony and consequent confusion.

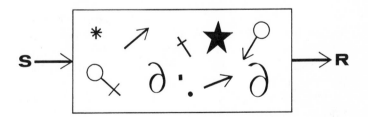

---

* Charles E. Osgood, "Behavior Theory and the Social Sciences," in Roland Young, ed., *Approaches to the Study of Politics* (Evanston: Northwestern University Press, 1958), pp. 217–244. The quotations are from pp. 219–220.

Notice here the extent to which even Osgood's historical account is concerned with "debunking previous speculations." Enter now the verification perspective. Osgood continues:

> In direct revulsion against this brand of theorizing, a group of American behaviorists around the turn of the century went to the other extreme, claiming that the psychologist was better off if he made no assumptions whatsoever about what went on in the little black box. The group included Weiss, Kantor, Watson, and somewhat later, Skinner. This viewpoint toward theory has come to be known as "empty organism psychology." According to this view, as shown below in diagrammatic fashion, there is absolutely nothing in the region between S and R, and what *is* there is none of the psychologists' business! This objective viewpoint was a healthy antidote for the loose mentalism which had preceded it, and it came to characterize American psychology. It led to increased emphasis on the details of accurately measuring stimuli and recording responses, to the establishment of dependable empirical laws relating input and output events, and to a general suspicion of unobservable explanatory devices, like "ideas," "purposes," and "feelings." But it also led psychologists to limit their interest to problems that could be handled in this simple mold—which automatically eliminated most phenomena of peculiarly human significance. The rat replaced the human as the standard subject in psychological research.

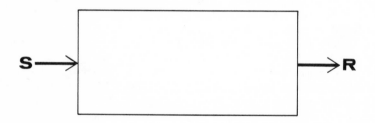

Osgood here states the impact of the verification perspective with great clarity. Initially there is an important question or set of questions in which men are interested. Speculation with respect to these questions is rejected in favor of greater methodological rigor. The result is that the questions are narrowed to fit the methods and the subject matter is itself radically altered. It is perhaps worth reiterating that this description is offered not by a hostile critic but by a prominent behavioral psychologist. One way of describing this development in psychology is to say that upon the scientific enterprise, that is, man's quest for knowledge—in this case about his own mind and brain—there was imposed both a mechanical model and an empiricist model. Let this stand as an assertion for the moment; we will shortly return to it and make it clearer.

Most contemporary *behavioral* scientists know about this chapter in psychology, eschew any connection with it, and regard it as ancient history. This is rather like a modern abstract painter denying any relationship to Picasso because "he used different colors than I do." Specialists always emphasize the differences between themselves and other specialists in the same field, but from a more general and truer perspective they may be nearly the same. It is quite true that these original assumptions have been modified and moderated to a considerable degree (the assumptions of the original Logical Positivists have been similarly altered), but their influence remains powerful and highly significant. Osgood, continuing from where we left him, makes the point quite sharply:

> Most contemporary behaviorists could be characterized as "frustrated empty-boxers." Armed with a minimum but effective set of principles— really a set of empirical generalizations gleaned from systematic observations of S-R functions—they set out to explain and predict behavior in general. It soon became obvious, even with rats as subjects, that something had to be postulated between S and R in order to explain our observations. But the contemporary behaviorist was more sophisticated about theory construction than his nineteenth-century forebearer and, furthermore, he was under constant critical pressure from his objectively oriented colleagues. Therefore he tried to put as little back into the box as possible, *i.e.*, postulate as few intervening variables as possible, and he tried to anchor these hypothesized constructs to antecedent and subsequent observables as firmly as he could.

Osgood proceeds from this to describe the ways in which various theoretical constructs were postulated to exist between stimulus and response, between input and output. While the "black box" is no longer thought of as being absolutely empty, Osgood quite rightly stresses the fact that the fundamental model of explanation—and I might add, the understanding of science which it implies—remains the same. Here is the mechanical, empirical model at its clearest and its cleanest. The human brain is treated as if it were a sort of machine which processes stimuli and turns them into responses according to certain rules. The intellectual objective is to discover what the rules are—they are understood to be logical equivalents to Newton's laws of nature. If in an experiment one begins with the stimulus X applied under certain conditions given that the general law is such-and-such, then we can predict that the response will be Y. If the response is indeed Y, then the general law tends to be confirmed, thus linking prediction (in a certain broad sense of the term) to explanation.

Notice also the empiricist element—one might alternatively say "the verification perspective"—in this model of explanation. Osgood speaks repeatedly of "observables," and of relating "observables" by general laws.

What he describes as "theory" and "theory construction" is simply the reverse side—so to speak—of the verification perspective. Thus, the verification perspective and the prediction-generalization model are—on the whole—two sides of the same coin. It is symptomatic of this line of thought that its practitioners abhor the use of the terms "mind," "mental," or "mental event" which presumably refer to an "unobservable." "We do not *know* anything about the 'mind,' " say the psychologists, "and neither does anyone else even though they may profess to do so." We know only about the general laws which relate stimulus to response.

But why? Why is psychology done this way? Why is this procedure employed? These are the questions which concern us here, and the answers could not be clearer than those given by Osgood himself. Not because "the free exercise of often brilliant intuition" could not be had by other means, but because of "parsimony" and the elimination of "confusion." There has been, Osgood says, a "refinement in the rules of procedure" and a "trend toward increased rigor." In short, psychology is done this way not primarily because of the demands of the subject matter but because of *science,* in order to make the procedures and the conclusions more *scientific.*

4

One cannot quarrel with the objectives of increased rigor or the elimination of confusion, and I do not for a moment do so. It is, however, clear beyond any doubt whatever that what Osgood here describes is a particular conception of science derived, as Kuhn says, "from the study of finished scientific achievements as these are recorded in the classics . . . and the textbooks." The classic case is the celestial mechanics of Newton with its laws of nature supported by correct "predictions" of celestial motion. There is nothing *prima facie* wrong with this particular conception of science. If there is something amiss here—and I shall argue that there is—it lies in the fact of not recognizing that this is but a *particular* conception of science.

We should, of course, not be surprised to discover that psychologists have adopted this policy. Our previous discussion explains pretty clearly why it happened. Scientific success until the beginning of the twentieth century—or at least the success of the "master" science of physics— rested upon this model and the early philosophers and historians of science duly recorded the successes and the explanatory policies involved in them. Psychologists, like political scientists, inherited a sort of "literary" tradition which they strove to make scientific, and they—as we might expect—picked up the tools of rigor and precision devised by the physicists and chemists.

But these tools were honed by the physicists and chemists to deal with the problems of physics and chemistry as they were understood in the eighteenth and nineteenth centuries. The physical scientists had certain problems and they devised tools to deal with just those problems. They did not invent "rigor" and "precision" in some abstract sense of those terms, but they were rigorous and precise with respect to particular intellectual puzzles. Neither—and this is of great importance—did they invent "scientific method" in some essential sense of that term; they devised *a* scientific method with physical and chemical problems in mind. Let your mind wander back over all of our earlier discussion of ideals of natural order and paradigms of explanation. Newton, Dalton, Maxwell, and Mach did not "just look at the facts" any more than Copernicus did —or than Copernicus could possibly have.

5

Another way of looking at our present problem is to say that the tools devised by the physical scientists involved a certain cost. I mean "cost" here in the sense of "cost in accurately describing reality." The prediction-generalization model of science—as we called it earlier—involves two sorts of cost to which I would like to call attention.

The first might be called the "three dimensional" cost. By this I mean that generalization requires the systematic distortion of the ordinary three dimensions of the objects which are generalized about. If one is to generalize—as in a classic case—about balls rolling down inclined planes, one must assume that all balls are perfect spheres—which of course, none of them is. Newton, in order to make the law of the inverse squares work, had to postulate a "contrapuntal mass," that is, he had to assume that the entire mass of say, Jupiter was concentrated in a single point. Nothing, of course, could be further from the truth. For a chemist to generalize about substances he must assume them to be absolutely pure, which, again, they never are.

"But," you may object, "aren't you making something out of nothing? What you say is, strictly speaking, quite true. But it really doesn't make any difference, because for all practical purposes the fact that the balls aren't quite spherical or the substances not quite pure doesn't really matter when you think of the advances made by treating them as if they were. So why call it a 'cost,' as if there were something wrong?"

But this is just the point! In *physics* and in *chemistry*—at least as these disciplines were understood in the nineteenth century—the cost is slight, and it matters hardly at all. This is what it means to say that the tools were devised by physicists and chemists to deal with the problems of physics and chemistry. The adoption of a particular ideal of natural order or paradigm of explanation always involves a cost, but in this case the investment is very small considering the dividends.

Think for a moment of what is called the Einsteinian revolution in physics. There are a great many things involved in Einstein's work, but Einstein in some sense did not show that Newton's laws of nature were wrong, only that their cost was too great in certain circumstances. Consider also the actual motion of the earth's natural satellite. While Newton's laws provide good reasons why the moon does not fly off into space or crash into the earth, they do not suffice when the problem is describing its precise motion—a matter which is in fact so complex that, so far as I know, it has never been totally resolved.

The second cost can be called the "fourth dimensional" cost. The fourth dimension is, of course, time, and the universal generalization is by definition timeless. Laws of nature as classically propounded by Newton and others are understood as applicable regardless of time and place. Time is a kind of neuter; it is obviously present, but it doesn't of itself produce anything. Here the previous discussion by Toulmin and Goodfield becomes relevant. The costs involved in the absence of a time dimension did not bother the classical physicist at all. They are so obscure in fact that it literally took an Einstein to find them. The cost is, however, a bit easier to see in chemistry with the discovery of isotopes and radioactive elements.

The point that I am trying to make with this discussion of cost is in some ways fairly obvious. Almost everyone who has ever thought at all about the nature of science recognizes that scientific theory involves an "abstraction" from reality. As a matter of fact this is a favorite argument of those who would make social science more scientific. "Scientists abstract from reality," they argue, "therefore we should also—not only *should* we, but we must!" As I suggested, there is a certain obvious sense in which this is true, and at this obvious level this true-as-far-as-it-goes assertion is often used as a justification for a sort of *carte blanche* orgy of abstraction. Thus, one often encounters assertions that what is called "formal theory" is somehow good in itself, and only very rarely a clear notion of good for what. This generalized conclusion about abstraction is possible only when one fails to realize that the cost involved for those particular scientists—who are understood as "*the* scientists"—was very slight. Think again about the objections of Kuhn, Toulmin, Hanson, and others to a conception of science drawn from the finished classical theories of the eighteenth and nineteenth centuries.

6

Suppose that you are captured by some sinister enemy of your country and that you are brainwashed in such a way that your entire life experience is erased, but your reasoning power is left intact. You are conscious; you are able to think, but your mind is totally blank—a sort of science-fiction *tabula rasa*. You have awakened in a completely empty cell. Suddenly a panel in one of the walls silently slides open revealing a small cubicle—let us say three feet by three feet by three feet—open on the side toward you so that you can see into it. In the cubicle is a mother duck sitting upon a nest containing several eggs. As you watch, the eggs hatch and the ducklings emerge. As soon as all of the ducklings are out of their shells, the panel closes leaving you looking at a blank wall. After a time the panel opens again and the same events take place. After the ducklings have emerged, the panel slides shut. This process is repeated over and over again, but the time intervals between openings vary randomly, the ducks are the same species but different individuals, sometimes the number of eggs is different, and, of course, the ducklings are in each case different individuals.

Remember that your mind is totally blank, even though your reasoning powers are perfectly good. You do not know what a duck is, what a nest is, what an egg is, or what a duckling is. You do not even know what a living thing is. Suppose that as the number of panel openings piles up, you recognize that the individual mother ducks are different, the individual nests are different, the number of eggs varies, and so on. You now are in a position to be able to abstract and generalize, your first inclination having been to try to describe the infinite details involved in even this limited experience. You might at first have tried to attend to the variations in the feathering of the ducks, the straws that made up the nests, and so forth. When you see that while the particular parts of your experience are each different they nonetheless can be categorized according to "*x*"ness (you do not, of course, know that this is called "mother-duckness"), "*y*"ness ("egg-ness"), and so on, you have *abstracted* and you are ready to *generalize*. Thus, you conclude "when x (a mother duck) sits upon y's (eggs) in a z (a nest), a's (ducklings) will emerge."

The openings continue and the same events take place. Totally satisfied, you begin in each case to predict—"the a's will emerge." And sure enough they do, and you realize that this series of correct predictions confirms your law. All goes happily and smoothly. But, alas, your captor (you do not know that you are a captive) who must be something of a philosopher of science, has a taste for the stories of Hans Christian Andersen.

He begins every now and then to put a swan's egg in the nest and when the shells break open, an "ugly duckling" emerges.

You, of course, notice the ugly duckling but what can you do about it in terms of your law? You do not know and cannot know that the ugly duckling's parents were swans or that he will grow up to be a swan—anymore than the ugly duckling himself knew it in Andersen's story. Your situation in the cell precludes any knowledge of growth or development. You see only a narrow slice taken from a growth or development sequence that occurs in a much larger (relative to the part of the process that you see) period of time. You have no notion of time—your situation is in effect timeless—so you can only generalize. The relevant parameters for understanding why a duckling is a duckling and a baby swan a baby swan are completely outside your grasp. The notion that someone is playing a trick on you is unavailable—you do not even know that there are "someones." Thus, to continue to reason—to keep on being "scientific" —you can only do one thing. You must abstract further. You must expand your notion of what an "*a*" (a duckling) is.

We can imagine that the jailer might put a baby mouse or kitten in a very fragile shell and thus oblige you to expand your variable further. He might train a dog to sit on the eggs or put remote-controlled toy soldiers in plastic shells which would break open at the proper time. And you, in your prison which is no longer merely physical but mental as well, would go on abstracting, expanding your conceptions of the variables in order to save the only process of reasoning available to you.

7

The story of the ugly duckling as I have related it here has a moral, just as surely as does the traditional version of Hans Christian Andersen. If you hold that the prediction-generalization model of science is *the* model of science—is science itself—then you are in an intellectual prison nearly as confining as the one I put you in a moment ago. There is, however, an important difference which has the effect of clouding vision and making the walls difficult to see. One who adheres to the prediction-generalization view in the real-world is *not* deprived of his life experience. Thus, he is able to mix the products of his life experience with his generalizations and the problems with generalization may cause him no significant practical difficulty. He will have no *practical* difficulty in the sense that he will not realize his intellectual confusion because he relies in all important matters on his ordinary life experience. Thus, a man who may firmly believe that the only really reliable knowledge consists of generalizations may be able to operate perfectly well even though he knows quite clearly that he does not possess any real generalizations. Let me now leave this observation to be picked up later and turn to a related aspect of the moral of the story.

A generalization in the prediction-generalization, scientific sense of the term cannot, in and of itself, *explain* change over time. To see precisely what I mean here requires some sharp focusing. I set the prison situation up in the way I did to show the sense in which the generalization model cuts a kind of cross-sectional slice out of reality. A generalization if it is to qualify as a law, as a universal generalization, by definition must be timeless in the sense that it applies at any time whatsoever. Think back to the prison, the brainwashing, the ducks, and the ugly duckling. Had you not been so imprisoned, brainwashed, and limited to the narrow slice of the life process of ducks and swans, you could have *explained* what occurred perfectly well. You would have simply said that the eggs were produced inside the mother-duck, fertilized by a drake, laid, kept warm for the gestation period, and finally the ducklings broke the shell and emerged into the world. The reason why the swan's egg hatched a baby swan could be disposed of with equal ease. Notice that the generalization produced in the prison situation does not *explain* what is happening. The test of this is that *in* the prison situation you were totally misled by the appearance of the "ugly duckling."

Now let me put you back into prison for a moment under the same circumstances, but let me change what happens in the cubicle. For the first fifty panel-openings the eggs are all duck eggs and only "a"s are produced. Then there is a pause in the openings. When they resume again, all the eggs are swan eggs and only baby swans are produced. The mother-ducks and the nests remain the same as before. Now it is quite clear that you could say that at time T_1 only "a"s were produced and at time T_2 only "b"s were produced, but you could not possibly explain the change. The reason for your inability is simply that I have structured the situation so that you can only generalize.

Suppose now a third prison situation. All conditions are the same except that there are now three phases, T_1, T_2, and T_3. T_1 consists as before of the hatching exclusively of duck eggs, T_2 reveals the mother, the nest, and a group of one-quarter grown ducklings. T_3 shows the mother, the nest, and a group of half-grown ducklings. You could perhaps under these circumstances be able to get an idea of what is happening. "First the eggs and small ducklings, then larger ducklings, then even larger ducklings—Eureka!" But notice how the original generalization fades into unimportance. If you did cling resolutely to the generalization, trying with great imagination to stretch the variables to include what you saw at T_2 and T_3, it would positively impede your seeing what was actually happening. To *explain* change over time requires a refocusing, a new kind of seeing. You must either drop the generalization altogether, or recognize that it is limited to a certain set of circumstances. If you cling to the idea of generalizing through T_2 and T_3 the cost is too great. It is so great in fact that you will not be able to explain or to understand what is happening.

I can readily imagine that a number of objections will be raised to my use of the various prison examples. In anticipation of this let me try to meet what I expect will be a typical one. "Well," it might be said, "your third prison example is a phony one. In the original case the generalization was about eggs hatching, but at T_2 and T_3 in the third situation there aren't any eggs hatching. So, obviously you wouldn't generalize from T_1 to T_2 to T_3." Yes, but remember that you do not know what an egg is at T_1 and if you drop the generalization at T_2 and T_3, what you are doing is recognizing that the generalization is limited and will not explain what is happening. I have nothing against generalizations as such. I only want to show that they have a limited use and that any conception of science which is built exclusively upon a notion of universal law is erroneous.*

8

There is a very easy and simple way to summarize the argument which I have been attempting to make. When one opts for the generalization model of science he seeks uniformities, he seeks that which does not change, he seeks sameness. Thus, the bent, the impetus of his theorizing is settled from the outset. Change as change is *theoretically* irrelevant. The uniformities, the "persistencies" are the standard. Changes are but deviations from the "perfection" described by the theory. Change *as* change cannot be *explained,* it can only be "accounted for" in terms of the uniformities. Thus, a theory which specifies that when *a* and *b* are combined *c* will result *cannot* explain the appearance of *d*. It can only "account for" *d* in the sense that either *a* or *b* or both were not present or present in the wrong proportions or something of the sort. In sum, then, on this analysis a change cannot be understood in and of itself, but only by reference to an alleged uniformity.

Let us return now to Osgood's description of psychology. Under the direction of an intellectual standard which could be satisfied only by generalizations, by laws of behavior, psychologists abstracted from the fantastic

---

* If you are interested in a full-dress philosophical treatment of the problem of explanation and the role of general laws in it, I recommend with enthusiasm the writings of Michael Scriven. See especially his critique of the popular position of Carl Gustav Hempel in "Truisms as the Grounds for Historical Explanations" in Patrick Gardiner, ed., *Theories of History* (New York: The Free Press of Glencoe, 1959), pp. 443–475.

number of variations present in human behavior and posited the empty box. Correlation of stimulus with response in terms of this model, Osgood reports, resulted in a few simple regularities. The attempt to produce a more general theory, however, demanded postulation of a more complicated relationship between stimuli and responses. Something had to be put into the empty box, although the demands of the intellectual standard were such as to keep the additional postulates to a minimum.

How are we to account for these additional postulations? The answer is simple. The original notion of an external, observable stimulus and response had to be expanded to include internal, unobservable, built-in stimuli and internal, unobservable built-in responses. As Osgood diagrams it:

$$ S \longrightarrow \rightarrow S \rightarrow S \rightarrow S \longrightarrow R \longrightarrow R \rightarrow \longrightarrow R $$

Notice the logic of this situation. The more complicated the phenomena with which you try accurately to deal, the more you must stretch your concepts—providing, of course, that you wish to retain them. In order to retain the generalization in the face of the ugly duckling, you must expand the notion of "duckling."

The costs of generalizing are always there. One must judge *in the particular context* whether or not they are worth paying. I must allow the psychologists in their own context to argue their costs for themselves and, as is apparent from the briefest exposure to the field, they hardly need an outside invitation to argue these questions.

9

I suggested some time ago that psychology is worth looking at for our purposes because psychologists have been among the principal science teachers for political scientists. This is true in two different but related ways. On the one hand, political science has put certain substantive psychological notions and particular techniques to work in understanding politics. On the other, political science has learned its conception of science itself in substantial part from psychologists and social psychologists.

While we are, of course, concentrating on the second aspect, I want for a moment to examine the second aspect a bit more accurately by citing an example from the first.

One of the most successful—perhaps the most successful—attempts to gain new knowledge relevant to politics has been the systematic study of voting behavior initiated some thirty years ago. While there have been several studies all differing from one another in a variety of ways, the use of psychological notions and techniques has been central throughout. Most of the studies have actually been done either wholly or in part by psychologists and social psychologists. What was being studied was the voting choice of particular voters and thus matters of individual psychology were the points at issue. These studies are clear applications of the generalization model applied in the context of behavioral psychology.

Let us focus upon the study of the 1948 presidential election made by Berelson, Lazarsfeld, and McPhee.* Here the response (the voting choice) was correlated with a variety of voter characteristics (religion, socioeconomic class, and so on,) serving as stimuli. Generalizations were then produced, for example, Catholics tend to vote Democratic; the higher the socioeconomic class, the more likely the Republican vote, and so on.

I call attention to this example because it illustrates in a concrete way two of the points that I tried to make earlier in more general terms. The first is the significance of what I called "life experience" as it relates to generalization. Consider the raw statistical conclusion that people of high socioeconomic class tend to vote Republican. In isolation this fact has no more significance than the possible generalization that people with widow's peaks vote Democratic. It has significance because our life experience tells us something about high socioeconomic class and something about voting Republican. The cost of generalizing here is at least partially met by life experience which in this case is experience with American politics.

This study also illustrates clearly the sense in which change cannot be explained in and of itself but can only be accounted for. The interviewers asked the respondents some time before the election how they intended to vote and then after the election how they did in fact vote. Something of a change—from Dewey to Truman—was recorded. The information produced by these questions could not, no matter how it might be manipulated, explain the change. To do this the authors relied again on life experience. In terms of the information itself and the generalizations derived from it, the best that could possibly be done would be to check the characteristics of the voters who changed and speculate from there.

---

* *Voting: A Study of Opinion Formation in a Presidential Campaign* (Chicago: University of Chicago Press, 1954).

Again, the cost of generalizing can be observed, but it seems to be worth paying. But notice, it is worth paying because a substantial quantity of "life experience" is around to carry the load and the generalizations are in effect limited and confined to a fairly clearly circumscribed context. Even so, the inability to *explain* in this scheme led the investigators to devise tests which got at the question of why such-and-such was done.

Thus, while "why questions" and life experience are used to bolster and explain the conclusions, the prediction-generalization conception of science continues to lie at the core of the enterprise. Consequently, these studies tend to reinforce the prediction-generalization model and it continues without challenge as *the* definition of science. This brings us to a more fundamental question—what effect does the prediction-generalization understanding of science have on the notion of a general science of politics?

5
A General Science
of Politics?

1

I first encountered the idea of building a science of politics as a college senior when I happened to visit the political science department of a prominent neighboring university. This department was in the throes of controversy over curriculum revision and the point at issue was the building of a general science of politics. I was at first mystified by the discussion, but when I entered graduate school the issue became a little clearer. At this point a question began to form in my mind which I often asked but to which I never received a satisfactory answer. The question was this: What would a general science of politics look like after you built it?

The best answer that I could get usually had to do with building blocks, that is, a great accumulation of small studies would eventually lead to a general science. And the general science would somehow involve "generalizations" and "the ability to predict." I would ask for a sample of one of these generalizations, a variation on the question, What would such a science *look like*? But I never really received a good answer. The typical reply was that "we are not ready yet to make any such generalizations, so I can't give you one," followed by an assurance that the generalizations would come in time—"when we get to be like physics.''

All of this seemed more or less plausible to me. I had not yet seen through the prediction-generalization model of science. I had had a rather substantial exposure to natural science, but it had been on the whole "textbook science" of the sort that Kuhn describes. I learned the steps of scientific method like everyone else. In addition the traditional texts of the philosophy of science had been part of my formal education. But through all of this discussion something continued to bother me and I would try occasionally to express it. "Are you sure," I would ask, "that the ultimate generalizations won't turn out to be extremely vague, general ideas that come very close to being truisms?" The answer to this was almost always a momentary blank stare followed by a changing of the subject.

I now realize in a way that I did not then, that what was at issue was the prediction-generalization model of science which we have gone to such lengths to explore. Let us now engage in a bit of speculation. What would a general science of politics *have* to look like if one tried to fit it into the prediction-generalization model? This is really not as difficult a task as it might at first glance seem. There are certain rules which we would beyond doubt have to follow. We would first of all have to abstract out of the great "blooming, buzzing confusion" of human experience something which we would call politics. Because the unbreakable rule is generalization we must inevitably deal with a uniformity, a particular uniformity which we would call politics. There is no choice here because generalizing about diversity is by definition impossible. Now the generalizations *must*—again by definition there is no choice, for we are talking about a *general* science of politics—apply in any place at any time. Therefore the concepts which make up the generalizations must be broad enough to fit "politics" at any time and at any place.

Finally, because we are concentrating on uniformities, particular changes *have* to be theoretically irrelevant. We can make the presence of change in general compatible with our theory, but no particular change can have meaning in and of itself. It can be dealt with *in the theory* only by reference to the uniformities. This can be done in one of two ways. Either the change can be declared outside the range of the theory and thus theoretically irrelevant to the generalizations which it contains. (This, for example, would be the case if a student put gum on the inclined plane and thus prevented the balls from rolling down.) Or, secondly, the change could be regarded as not a change at all, but only a special instance of the uniformity. In this case the change is simply swallowed by the theory and, of course, is not and cannot be explained in its own terms.

We can expect, I think, that our general theorist confined to the prediction-generalization model will be in a situation quite like that of the prisoner whom we described earlier. Confronted as he surely will be by diversities many, many times greater than those we put into the prison situation, he will nonetheless be forced—by the fact that he is restricted to the prediction-generalization model—to expand his concepts and to make his uniformities more and more general as he encounters greater and greater diversity. If we had faced our prisoner with comparable diversity—not stopping with mice, kittens, and toy soldiers but replacing the various original objects with bears, tricycles, microscopes, brassieres, and so on—he would have been driven to a generalization as informative as "When the panel opens, there will be things in the cubicle."

2

We could, I am sure, continue this speculation almost indefinitely, but I trust that there is no need to develop the point by further hypothetical example. We must now face an actual attempt at general theorizing dominated by the prediction-generalization model. Let us, therefore, look in a general way at the theoretical writings of the University of Chicago political scientist David Easton.* One must at the outset at least, regard it as an open question whether or not a particular theorist is in fact dominated by the prediction-generalization model. With respect to Easton it seems to me that there can be little doubt that this is the case. In fact one cannot even follow his argument, make sense of his transitions, let alone accept his conclusions, unless one at least understands the prediction-generalization conception of science. I shall attempt to summarize his principal ideas and in the course of the summary the role of the prediction-generalization model will be perfectly clear.

In a good many ways there is no need to argue about Easton's perspective because he is on the whole quite explicit about what he is doing. He suggests that it is "useful" to define politics as the authoritative allocation of values for a society. Thus, the definition is not an essentialist one—it does not purport to say what politics "really is," but only that this conception is useful for scientific understanding. From the outset then he is abstracting and this abstraction is justified as proper scientific

---

* These writings are principally contained in three volumes, *The Political System* (New York: Alfred A. Knopf, 1954), *A Framework for Political Analysis* (Englewood Cliffs, N.J.: Prentice-Hall, 1965), and *A Systems Analysis of Political Life* (New York: John Wiley & Sons, 1965).

practice—and so it is. But notice that the particular definition chosen—the way of abstraction—is one that seeks uniformity. The notion of the authoritative allocation of values is intended to apply to all societies.

Easton ties this definition to the idea of a system. System here is a notion derived from biology and it is intended to suggest an interrelated set of variables, interrelated in such a way that when one changes all the others are affected. Physiology, for example, broke down the organism into the respiratory system, reproductive system, and so on. Both as a matter of history and as a matter of logic this notion of system is biology's version of Newton's mechanics. It is, in short, a sort of machine that is alive. The solar system, for example, is different from the digestive system in at least two important ways. The solar system has an obvious natural boundary, the digestive system does not and thus the boundary must be settled theoretically in terms of some notion of "usefulness." Secondly, the solar system is static in a way that the digestive system is not. The digestive system processes various substances which enter it (cross its boundary) and it must therefore adapt itself to different substances.

The idea of system in physiology, while different from a purely mechanical system, is nonetheless Newtonian in its origins and a clear case of the application of the prediction-generalization model to physiological subject matter. Physiologists were and are able to generalize and predict in its terms about reproductive systems, digestive systems, and so forth. The intellectual objective is to be able to describe digestive, respiratory, reproductive, and so on, systems *in general,* and not simply with reference to a particular animal or species.

Easton's system is close to that of physiology. The political system receives "inputs" and converts them into "outputs." Thus, Easton's analogy is not a simple mechanical device, but something rather more complicated. He suggests that:

> . . . We might compare a political system to a huge and complex factory in which raw material in the form of wants, in our generalized sense, are taken in, worked upon and transformed into a primary product called demands. Some few of these are then found to be appropriate for additional processing through a variety of intermediary operations until they are ready to be converted into finished products or outputs, called binding decisions and their implementing actions. These finished products leave the system to act upon society as a whole, with consequences that may make themselves felt subsequently through the generation of additional demands that seek entry into the system. This forms a closed-loop process that I shall later characterize as feedback.
>
> Alternatively, [Easton continues], we may visualize a political system as a

gigantic communications network into which information in the form of demands is flowing and out of which a different kind of information we call a decision emerges. If such an output is to be possible, there must be various intermediary processes the consequences of which are to permit passage of, winnow out, combine and recombine the incoming messages so as to mold them into a number and kind that can be conveniently managed by the decision-makers. Our task is to examine and understand the way in which the conversion process moves this information along the network from the point of entry to exit, regulating any possible stress on the system that might emerge.*

Even these analogies, Easton assures us, are oversimplified, because the political system is much more complicated than either a factory or communications network. Notice, however, the logic of the analogies. While they are more complicated and we may take Easton's word that the political system itself is even more complex, the logic is the same as the old *S,* empty or partially-filled box, *R* model described by Osgood. We continue to deal with the same logical perspective as that of the early behavioral psychologists and it is present for the same reasons, namely, the power of a particular conception of science.

It should now be clear that Easton seeks uniformities, that he seeks general categories which will deal with politics at any time and at any place. There can be no doubt about this because Easton repeatedly and explicitly states that this is his intention. We can also see the *way* in which he intends to generalize, we can see the particular kinds of uniformities that he thinks are important. Let us skip any minor quarrels that we might have with the idea of the system or with Easton's use of it. We must examine something more important. Let us see what Easton does with his "ugly ducklings."

3

How much does it cost to abstract in this way? We can, I think, accept without much hesitation the general idea of politics as a system. It is a plausible analogy, it focuses on the interconnection of diverse factors, it does not do unreasonable violence to what we ordinarily think of when we hear the term "politics." The notion has undoubtedly a certain merit as an expository device. If we want to describe what, say, the government of the United States does day-by-day, the system idea provides a certain

---

* *A Systems Analysis of Political Life,* p. 72.

clarity which is helpful. But how much good does it do as a general theory? What happens when one tries to make it into something much more than an expository device, as Easton surely does?

Let us go through this exercise in general theory building in a necessarily attenuated but nonetheless step-by-step fashion. We need, first of all, to recognize that Easton begins against a background of what he would call traditional, descriptive, "common-sense" political science. So let us talk traditional common sense for a moment. If someone asked "What is government and politics in the United States all about?" we could give a certain kind of common-sense reply. Well, we might say, first there was the English settlement in North America and then the revolution and the Articles of Confederation and then finally the writing of the Constitution in 1789. Now this constitution created certain institutions—the presidency, the congress, the Supreme Court—and people in the United States let those who man the institutions know what they want and the officials make legally binding decisions which everybody except the lawbreaker follows.

The typical American government textbook tells this story in much greater detail, often quite skillfully. If this is pushed a bit further and one asks "What is government and politics all about in a general way?" a reasonable answer, still at the common-sense level, would be something like the following: "When people live in a society they have certain needs. They express these needs in a variety of ways. The governmental institutions receive these requests, decide on a policy, and enforce it by law." American political scientists have been making detailed studies illuminating this description and matters related to it for years. It is against this background that Easton with his inclination for scientific generalization appears.

Now it is clear enough that this common-sense account, even in the more general version, is insufficiently general to qualify as scientific theory on the prediction-generalization model. It is descriptive only of a particular kind of political operation, one of which the United States happens to be a very good illustration. Modern institutions, for example, are temporal, man-made things and they are not always present. Thus, in keeping with the dictates of science one must look for the real uniformities of which modern institutions are only an example. For Easton, therefore, political life should be seen as an "adaptive, self-regulating, and self-transforming system of behavior" which responds through the process of feedback to internal stresses as well as to those which originate in the environment in which it finds itself. Now, since we are talking not about any particular objects but a completely general "system of behavior" we must exclude any notion of traditional structures like legislatures, parties, or, for that matter, persons as biological entities. We concentrate instead on the set of relevant interactions among members of the system.

But, we may properly ask, what defines the "relevant"? The relevant interactions are political interactions defined as such by virtue of the fact that they are "predominantly oriented toward the authoritative allocation of values for a society." We have, thus, reached the point where we are able to say that the political system is an "adaptive, self-regulating, and self-transforming system of behavior" which is made up of "interactions" "predominantly oriented toward the authoritative allocation of values for a society." Notice that a political system is a political system because of what it does—it authoritatively allocates values for a society. A scientific theory, of course, is more than a definition. Thus, a general science of politics becomes in part at least the study of *how* the political system authoritatively allocates values. Enter now the factory image. The political system may be seen as a "vast conversion process" which processes inputs of demands and system support, turning them into outputs which allow the system to "persist" in its basic characteristics.

Let us concentrate our attention for a moment on the word "persist." The notion of the persistence of the system is central not only to Easton's theory but to his exposition as well. Everything hinges on the system persisting. What function does the idea of persistence have in the theory? It seems perhaps to answer the question, Why should the system do what it does? It seems to supply the motive force, the energy of the system. Suppose we ask what makes the system authoritatively allocate values? Why does it convert inputs into outputs? Easton never really faces this kind of question. He simply asserts that this is what the system does. He does, however, repeatedly say that the production of outputs allows the system to persist. The idea of persistence, so far as Easton's exposition is concerned, is simply carried into the theory of the political system with the general theory of systems derived from biology. In a biological context the idea of an organic system has a rather clear role for the notion of persistence—it is literally a question of life or death.

The waters, I am afraid we must say, are beginning to get muddy. Easton slides over the question of precisely what role "persistence" plays in the political system and it is clearly to his rhetorical advantage to do so. We cannot, however, let this matter slide because it is central to the validity of the whole enterprise. What Easton does, it seems to me, is implicitly to bring over the notion of persistence from the theory of the organic system. In the case of an organic system it may be said from a theoretical point of view that the various parts of the organism (the system) function as they do so that the system may persist. Thus, the life of an organism is theoretically explained in terms of a system that functions so that the organism can persist. Easton frequently speaks of the vital "life processes" of the system, and in the case of the organism if a vital process fails to function properly the system will not persist, that is, the organism will die.

Remember now that Easton, because of his need for uniformity and generality, specifically excludes people as biological entities from the political system. The system is emphatically not a group of people, it is a "system of behavior," the set of relevant interactions. Therefore, we might suppose, if the political system is to fail to persist, it is the system of behavior that dies. Recall now that according to the theory of the organic system the parts function as they do so that the system will be able to persist. If we bring this notion over into Easton's account *in an explicit way* we must conclude that the political system authoritatively allocates values in the particular ways that it does (turns inputs into outputs, receives feedback, and so forth) in order to allow the system to persist. If what I have said here is correct, a very interesting thing happens to Easton's theory. A theoretical system of behavior which is *defined* as a system which authoritatively allocates values—not randomly or for the hell of it—but in order to allow the system to persist *could not possibly fail to persist*. Remember that the system is an abstraction, a theoretical construct. Death for a system *defined as a system that persists* would be self-contradictory.

4

Let us be clear about what I am arguing here. Biologists looking at the life processes of an organism see as an obvious empirical matter that the various parts—organs, tissues, cells—function interdependently in what appears to be an ordered way. One way of accounting for this ordered functioning would be to accept a teleological principle. It is possible to argue that blood circulates, the eyes blink, and so on, for a certain purpose, and that every organic process is directed to a certain end, and that there is an overriding end that might be described as the health and natural development of the organism. The notion of teleology, of course, pushes one into a philosophical thicket. It suggests a general purpose in the universe, it raises the Thomistic and Aristotelian notions of natural law and so on.

In order to avoid this difficulty biologists found it useful to develop the notion of system and to use the concept of the persistence of the system to avoid the idea of a *telos* or natural end. Strictly speaking, of course, the system of the organism is a theoretical construct, but the "fit" with the empirical reality is very close indeed. One could, I think, argue that, speaking strictly, the system—a theoretical construct—cannot itself die, but the actual physical death of the organism is not only a possibility but an eventual certainty. Thus, the notion of "failing to persist" has a very concrete referent. What we have here is another instance of abstracting where the cost is slight and well worth paying.

As I suggested a moment ago, Easton does not face up squarely to the role that "persistence" plays in systems theory. Apparently he just brings it into his political discussion because it is a regular part of systems theory. Consider the following discussion which serves to introduce the idea of persistence in *A Framework for Political Analysis.*

> For purposes of what we might call an allocative theory (the form of theory implicit in traditional political research) as against a system-coping and persistence kind of theory, we would be predisposed to ask other kinds of questions. How do political systems operate? How do they in fact allocate values? . . .
>
> For systems analysis, however, certain key problems logically come before these. They concern the conditions under which such allocative processes and structures themselves persist. These are the life processes of any and all systems.
>
> At the theoretical level, for example, it is as though from an interest in personality systems we were to set aside inquiry into specific personality types and the behavior of individuals characterized by such types. Our main task would be to probe into the general processes common to all personality structures, through which threats imposed upon the integrity of the system might be handled. In a comparable way, *systems analysis impels us to direct our attention to the very life processes of political systems rather than to the specific structures or processes that make a particular kind of regime viable.**

Easton leaves us with little doubt that he is directed by theoretical considerations. Systems theory "impels" us to focus upon the "very life processes of political systems" and to see how it is that these "life processes" persist. But what are these "life processes"? They are the "allocative processes and structures"—not the "specific structures or processes that make a particular kind of regime viable"—but the "allocative processes and structures" in general. Thus, the life processes are whatever it is in any particular case that "authoritatively allocates values for the society." Easton's definitions of politics on the one hand and of the "life processes" of the political system on the other are identical.** Consequently, Easton

---

* *A Framework for Political Analysis,* pp. 78–79. (Italics added.)
** *A Framework for Political Analysis,* pp. 95–96.

is obliged to say—as indeed he does, although not in so many words—that, in his sense of the term "persist," a political system fails to persist when and only when his definition of politics fails to apply. But—and this is all important—his definition of politics is meant to be general, to apply to any case whatsoever. We are led, therefore, to the following conclusions. The political system fails to persist when there is no longer any politics, or politics fails to persist when it fails to persist. The political system persists when there is politics, or politics persists when it persists.

5

We must recognize, I think, that there is something seductive about the notion of "persistence" for anyone who is trying to understand politics. Anyone with such an interest would immediately see the significance of the failure or the continued success of a particular regime. How is it that the parliamentary system in Great Britain has managed to persist through thick and thin for many hundreds of years? Why did the *Ancien Régime* in France fall? Why did German democracy of the 1920s turn into the New Order while American democracy of the 1920s turned only to the New Deal? All of these are pertinent questions and good answers to them would surely constitute political understanding of an important sort. Because all of these instances and many others which could as easily be suggested involve matters of "persistence" or "failure to persist," Easton's use of these terms has a certain appeal. But what, we must ask, can Easton's theory in and of itself tell us about any of these cases?

A sensible way to get at this question would be to focus on certain crucial dates. In the British case one might pick 1643, 1688, 1832, and 1945. In the German and American instances one could probably be quite precise and settle on January 30, 1933 and March 4, 1933. Let us focus here on what is an extreme case of "failure to persist"—France in 1789. There is no need to go into the details of the French Revolution here. All we need to say is that before 1789 a certain kind of political order existed in France and that sometime after 1789 a very different sort of political order came into existence. We may say then that we have situation A, the regime of Louis XVI and situation B, the First Republic. Easton's completely general definition of the political system, of course, includes both A *and* B. There was "an authoritative allocation of values" for France both in the time of Louis XVI and in the time of the First Republic, and we must conclude therefore that the "very life processes" of the French political system *persisted*—not, it must be said, because of anything that happened in France but because Easton's definition continues to apply.

Recall now the remarks made earlier about systems theory as used by biologists to understand the life processes of an organism. The system, I suggested, as a *theoretical construct* cannot die in the strict sense, but the organism itself will certainly die. Thus, the "fit" is close, the abstraction is not terribly costly. When Easton talks about a political system failing to persist his principal examples are earthquakes and epidemics*— if the population is destroyed, if the organisms die, then the definition will no longer apply. But, apparently, as long as any semblance of human interaction remains no matter what its form, the definition still applies. How costly is an abstraction, allegedly relevant to politics, that leads us to conclude that the French Revolution, the coming of the New Deal, the Labour victory in 1945, Hitler's taking power, and indeed, I suppose, the German conquest of Poland are *all* instances of persistence?

6

In all of the discussion of Easton's writings up to this point, we have been looking at them, so to speak, from the "outside," as an instance of the prediction-generalization model in action. If we are to get the full lesson from Easton's theory, we must, I think, shift our focus and try to look at it from the inside. Professor Philip E. Converse of the University of Michigan, himself a distinguished contributor to contemporary political science, reviewed Easton's *A Framework for Political Analysis* in the *American Political Science Review*.** The review, I think it is fair to say, is a sympathetic one, referring to the book as, among other things, "this important theoretical contribution." I should like to quote successive paragraphs from Converse's review interjecting comments where they are relevant. "The remarkable capacity," Converse begins, "of the life processes of politics to persist amid endless social and political stress provides the central theme from which David Easton builds *A Framework for Political Analysis*." He continues by summarizing the theory very much as I did earlier. Converse comes then more directly to the question of persistence.

* *A Framework for Political Analysis*, p. 82.
** Converse, *American Political Science Review* (December, 1965, vol. LIX, no. 4), pp. 1001–1002.

It is this persistence in the face of stressful change that Easton highlights as a miracle of the commonplace, warranting attention as the primary goal of political analysis. "Persistence" here is not to be read as stability, maintenance of specific structures, or even "equilibrium" as commonly construed. The reference is to a more fundamental capacity of political systems to allocate values authoritatively. This capacity may endure through all sorts of change in specific structures or even basic forms of government, as in the case of a democracy turned into a dictatorship. Transformations at these more specific levels are seen as either irrelevant to the question of persistence in the form that Easton intends to pursue it, or as indicative of ways in which the system copes with adversity in order to preserve the vital processes of its political life.

As Converse suggests, Easton cannot allow persistence to mean simple stability, maintenance of particular structures, or equilibrium because not all political systems exhibit these characteristics at all times and because such notions tend to involve a sort of "status quo" bias—as critics of such "static" theories have been more than willing to point out. Thus, the necessity of generalizing drives Easton to a broader conception which includes the possibility of change. Recall now my remarks several pages back when we were discussing the matter of allowing for change in a prediction-generalization model. I suggested then that any *particular* change could be dealt with in one of two ways: either it could be disregarded as irrelevant or "swallowed" as a special case of the uniformity. Now, reread the final sentence in Converse's paragraph above. Converse continues:

> By moving to this level of abstraction, the author outflanks the most familiar criticisms of other general theories with similar roots in functionalism. Change, be it even of broad and dramatic dimensions, secular as well as compensatory, is in no sense alien to Easton's conception. To remain viable, societies must preserve the fundamental life processes of politics, but this may be as well-accomplished at some junctures through sweeping change in political forms as by any strain toward the *status quo*.

It is perfectly true, of course, that change *in general* is not alien to Easton's conception but, as we have just seen, any *particular* change must either be disregarded or swallowed—again because of the abstractness of the theory. We can in no sense then regard Easton's theory as a theory of political change—as a theory which answers questions concerning why any particular political change occurred. As Easton himself says, "My approach to the analysis of political systems will not help us to understand why any specific policies are adopted by the politically relevant members in a system." * He is even contemptuous of the idea of a theory of change: "Although social science has recently and suddenly become enamored of change and a tidal wave of theories of change threatens to engulf us, it has at least opened our eyes to the fact that any general theory, if it is even minimally adequate, must be able to handle change as easily as it does stability. But the truth is that in the elaboration of the initial fundamental categories of an analysis, there is no need for special concepts to study change. Indeed to introduce them would be a sign of weakness and a disjunction in the theory, not one of strength and integration." ** It is interesting to notice the wording here. A "general" theory must be able to "handle" change—not "deal with," "account for," or, heaven forbid, "explain," but "handle." What "handle" means in this context is that the theory must supply terminology capable of describing *that* such-and-such a change occurred, but, by Easton's own admission, it can *explain* no political change whatsoever. Thus, Easton's analysis stands in sharp *logical* contrast with a theory like that of Marx, which, whether factually true or false, is logically suited to explaining *why* the French Revolution, for example, occurred. Pause to reflect on the old question of "cost."

But let us return to Converse.

This hospitality to change opens the way to the possibility of a dynamic analysis, and represents for me one of the most exciting aspects of the book. At the same stroke, the necessary shift in level of abstraction threatens to rob the question of persistence of its savor for me, although clearly it does not do so for the author. The nub of the problem is the negative case, or what it means for the political system in this most overarching sense to *fail to persist.*

At this point, the "general systems" analogy is at its weakest. Biological organisms brush with the limits of survival upon more or less frequent occasions and the gravity of these approaches is certified by the fact that in Lord Keyne's long run, the survival probability of the organism is an unrelenting zero. But what do we have in mind when we speak of a political system "succumbing," once we have set aside any question of the life and death of particular modes of value-allocation? If the political system is constituted by interactions oriented toward the authoritative allocation of values within whatever geographic scope the society can be said to function, and is considered to persist as long as such interactions persist, then it is hard for me to see what will stop them short of the catastrophe that wipes out the population engaging in them. Indeed, earthquake and epidemic are cited first in

---

* *A Framework for Political Analysis,* p. 89.
** *A Framework for Political Analysis,* p. 106.

those passages wherein Easton sets out to describe negative cases, although these scarcely strike me as theoretically exciting. Once again in vivid contrast to the case of the organism, failure to survive through such causes is extremely rare and hinges on variables exogenous to almost any study of political process.*

As all of our previous discussion surely indicates, Converse has just cause for questioning the vagueness of the notion of persistence as Easton presents it. What is at issue here—and Converse sees this at least partially, although Easton apparently does not see it at all—is that there are far too many "ugly ducklings," not to mention kittens, mice, and toy soldiers, in the world of politics to be contained by a generalization that is anything more than a truism. The fact is, of course, that no system can fail to persist if the system is *defined* as one that persists. Easton, quite ingenuously it seems, struggles with this problem, but it is not a little amusing to watch the self-proclaimed prophet of the new wave of "empirical theory" try, like the metaphysicians of old, to force an infinite number of angels onto the head of a pin. Some of his attempts are almost, but not quite, plausible. Converse takes them up:

> The few other negative cases hinted at are of much greater potential interest, although they fail to satisfy my curiosity because either there was no failure to persist, or else the instances seem rather to involve changes in more specific forms of authoritative allocation. Thus, for example, the author considers the Congo to have come as close as any state in the twentieth century to the brink of survival, although he also deems that its political system managed to persist. Now the brush with survival at a certain level in this instance would be plain to all, whatever the theoretical viewpoint. Yet attempting to assess the world from within the Easton framework, I would have seen the Congo as representing more the kind of transition between types of government (centralized, decentralized) that the author in every other context sets apart from his abstract concerns with the underlying, "life processes" of politics. Surely if violence is but one variant on the generic processes whereby authoritative allocation occurs, as was argued in *The Political System,*** then the sense in which such vital processes were near extinction even in the Congo would escape me.†

Again, I think that we must say that Converse is quite right about this. Try as we might to discover a distinctly *political* instance of "failure to persist" in Easton's terms, such an example cannot be found because it has been defined out of existence from the very beginning. What would

* *American Political Science Review* (December, 1965, vol. LIX, no. 4), pp. 1001–1002.
** *The Political System.*
† *American Political Science Review,* p. 1002.

it be like to find a society in which values were not *in some sense* authoritatively allocated? Easton, by accepting the prediction-generalization model as his intellectual standard, pushes himself into a dilemma from which there is no escape. One could fairly easily supply instances of failure to persist in a political sense if we were willing to indicate precise desiderata for "value," "authoritative," or "allocate," but to do so these desiderata would have to be arbitrary or "value-laden." For example, it could be said that the values were not being allocated in a genuinely "authoritative" way unless, say, consent of the governed were involved. Likewise, one might say of a particular instance that it wasn't *really* a value or it wasn't *really* being allocated for some reason or other. But Easton's science, of course, will not let him do anything like this. The alternative, which, it seems, Easton takes, is to stretch the concepts (to abstract further) to include any conceivable instance of the authoritative allocation of values, and this, of course, results in the emptiness which makes Converse so uneasy.

What lesson should we learn from this examination of Easton's argument? It might be suggested that Easton simply does the job of general theory building badly. The proper task, if this were all that were involved, would then be to find better definitions, to manufacture better concepts, in short to do Easton's job better. We would, I think, be quite wrong to accept this conclusion. The fact is that any attempt to make absolutely general statements about politics, statements intended to apply regardless of time and place, will end in truism or something very close to it. From this point of view the dilemma will always be there. Either limit the generality arbitrarily or in accordance with a "value" or the generality will result in vacuity.

The problem is quite comparable to that involved in making absolutely general statements about science. No statement that is not so broad as to be nearly meaningless will cover all of science, as we saw earlier. The only way out is to make an arbitrary limitation, but to make such a limitation is no longer to be general. And there we are! Toulmin, Kuhn, Hanson, and the rest object to the earlier abstract conception of science because they see that the abstraction costs too much. Art is another comparable case which, however, for reasons that in another context might be worth speculating about, is rather easier to see. General statements about art would inevitably either be empty—"art is creativity"—or subject to the charge of being limited either arbitrarily or by a matter of taste.

Easton really performs a considerable service for students of politics, although certainly not the one he intended to perform. He painstakingly creates the *reductio ad absurdum* for the idea of an absolutely general, any-time-any-place, theory of politics. By showing us that a political system, construed in an absolutely general way, fails to persist only in those cases where it fails to persist, Easton convinces us of what no mere critic —because he would lack the necessary persuasive and psychological leverage—of the idea of a general theory of politics could possibly convince us of, namely, that the enterprise is futile.

PART TWO

The Evolutionary-Developmental Paradigm: A Speculation

6
On Taking
Time Seriously

It is necessary, I think, to try to make the argument of the preceding chapter as sharp and pointed as possible before moving on to other matters. My remarks about generalization and abstraction do not by any means constitute an attack on generalization and abstraction *per se*. On the contrary, generalization and abstraction are absolutely necessary for most forms of human communication. Presenting a reasonably complete account of the uses of generalization and abstraction would be the task of a book at least as long as this one. Let me, for present purposes, point only to a particular but significant distinction which, it seems to me, helps to make sense of my earlier remarks.

An abstract generalization is always a sort of shorthand description of some aspect of reality, or at least it is intended to be such a description— a particular generalization might be false and would, thus, hardly qualify as a description. If we say, for example, "Americans are liberal individualists," we are abstracting, we are generalizing, and above all, we are describing a complex reality in a shorthand way. The adequacy of a generalization of this sort has to do with the extent of its "fit" with reality, its appropriateness given certain purposes, and so on—and all of this has almost nothing to do with our discussion in the last chapter. We have no quarrel with descriptive generalizations nor for that matter anything unusual to say about them.

Notice now that the descriptive generalization can sometimes be used for explanatory purposes. If, for example, someone asks, Why is there no significant socialist party in America? we can explain in part by saying "Americans are liberal individualists." Here we could always ask for more by way of explanation—Why are Americans liberal individualists?

Suppose, however, the situation in which a generalization is offered as the ultimate explanation—"this is as far as you can go because the generalization describes the fact that nature just *is* this way." No further explanation can reasonably be asked for, things simply are this way! What we are talking about here are, of course, laws of nature in Newton's sense. What is at issue is emphatically not the idea of generalizing as such, but the generalization *model* of science which implies (1) that universal any-time-any-place laws are *the* goal of science, and (2) that any sort of generalization is good science just because it is a generalization. My argument in the previous chapter is simply that any attempt to develop an explanatory theory of politics on this model necessarily, given the diversity of the subject matter which must be dealt with, ends in vacuity. From this point of view then, Easton's theory is not important in itself. It is but an example of what happens when such an attempt is made.

Saying all this, however, does not mean that Easton's theory is devoid of descriptive content. It might be—in the sense that calling political orders systems might be false—but this would be a different order of observation. Easton, for all of his professed empiricism, is really in very much the same position as the classical metaphysician. He supplies us with a "linguistic recipe" in terms of which observations about politics can be stated.

It is clear enough, I think, that one could not *in terms of Easton's theory* explain why any political order failed to persist because all political orders are defined as systems that *do* persist. By the same token it would be impossible to explain why any particular political order did manage to persist because all political orders are by definition persisting systems. But this is not to argue that one cannot describe, say, the political events in Russia from 1915 to 1920 in Easton's terminology. It is not absurd to say that there was something wrong with Czar Nicholas's feedback loops—although this might add a measure of levity to the situation which some might find undesirable. It is this descriptive residue, this "linguistic recipe," that gives Easton's notions whatever persuasive power they have. And this power, I think, is not negligible. As expository devices the system idea and the "flow model" when applied, for example, to politics in contemporary Britain or America have considerable merit.

The adoption of a particular vocabulary, of a linguistic recipe—while it may result in a sharpening of focus with respect to particular aspects of a situation—is quite likely to distort other aspects or fail to show them altogether. The inclination to generality, however, demands that all aspects be included, and therefore—just as advocates of the omnipresence of God are driven, when confronted with instances of evil, to defenses like "God works in mysterious ways"—the proponent of the universality of the system must show the system persisting when anyone else would say that it has failed, must call things "supportive" when anyone else would call them "destructive,"—and so on.

If it is the case—I think this would be difficult to argue against—that any linguistic recipe distorts or ignores some aspects of the subject matter with which it purports to deal, we may reasonably expect to find some things distorted or omitted in Easton's theory of the political system. The test, of course, from a descriptive point of view is whether or not the aspects omitted or distorted are important.

We can perhaps get at this question by looking for a moment at the basic characteristics of what we have come to call systems. Without getting too elaborate about it we can say that what is meant by calling something a system is that it is a sort of self-regulating mechanism. Thus, an example often used as an introduction to the idea of system is the thermostat. Of course, any picture of political systems would be much more complicated than a thermostat, but such a picture, as Easton's elaborate presentation illustrates, would continue to be of a self-regulating mechanism.

If we grant, as we have earlier, that it makes considerable sense to think of the government of the contemporary United States as a system, as a kind of elaborate thermostat, then suppose we ask, How did it get that way? This is rather like asking about a particular thermostat, who designed it when for what purpose? Now it would never occur to anybody to suppose that these questions should or could be answered in terms of the theory of self-regulating mechanisms which explains how the thermostat works. In terms of the theory of the thermostat the questions would not even be relevant. But by a similar logic the theory of the political system tends to *make* questions such as: Who designed it? or How did it get this way? irrelevant.

But are such questions irrelevant to the empirical world of politics? I hardly think so. There is a powerful case, it seems to me, for arguing that one of the primary reasons why the American political order operates as a more or less self-regulating mechanism is because it was explicitly designed that way.* Madison and the Founding Fathers did not, of course, know about systems theory, but they did know—as historical research has demonstrated beyond any doubt—about mechanisms, and they explicitly and purposely designed the American constitutional structure with the mechanical model in mind.

Perhaps the most influential school of thought in the field of the interpretation of American democracy has for many years contended that the Founding Fathers sought to set up a kind of oligarchy, an aristocracy of intellect and property. If this is so, one might reply, why did they bother with elections, majority rule, and the elaborate system of checks and balances? Does it then follow that they really *were* democrats in some populist "the people shall rule themselves" sense after all? There has been a good deal of close and careful reasoning about these questions in the context of what is called democratic theory.** Is it not, however, quite obvious—although so far as I know it has never been explicitly pointed out—that the Founding Fathers were trying principally to accomplish two important and quite compatible objectives? On the one hand, they were trying to ensure rule by the most capable. On the other, in their distinctly mechanistic way they were trying to create a social device which would allow for and help to create what systems theory would call feedback.

I do not intend to suggest by these remarks anything so banal as the notion that contemporary systems theorists are simply rewriting *The Federalist*. I do, however, explicitly intend to say that the framers saw the American constitutional structure as a device for processing demands for social change and that they saw the necessity of creating access points and of setting "power against power" to this end. What they said, what they thought, and what they did was clearly set in the context of eighteenth-century scientific understanding, political inclination, and moral passion; but the framing of the Constitution was a conscious act of political invention and not simply the automatic adjustment of a self-regulating social system. From the point of view of the theory of the universal political system, however, it is very difficult—if not logically impossible—to see a group of men explicitly choosing to *create* a political

---

* Consider "system" as a normative concept. Perhaps the trouble in many contemporary societies is not a weak system (in Easton's sense of system) but no system at all in Madison's sense. Perhaps the objective of nation builders ought to be creating a system where there was none before.

** See for example, Robert A. Dahl, *A Preface to Democratic Theory* (Chicago: University of Chicago Press, 1956).

system with central control, receptors, feedback loops, and so forth where none existed before. But in a very important sense this is what the men of 1789 in fact did.

Now, of course, it is possible to argue—and I suppose that Easton would—that the "very life processes" of a system inaugurated by English colonization persisted through the revolution, the Articles of Confederation, to the Constitution. To argue this way, however, is simply to flex metaphysical muscles, to bend and stretch a linguistic recipe, and thus to turn a conscious and deliberate choice into an automatic readjustment. And here I think is a distortion which must give us pause. Think back now to our discussion in the previous chapter and notice that when the idea of persistence is in effect used as the energizing force of the political system—that the system is "trying" to persist in what "makes" it authoritatively allocate values—this automatic operation of the system is in fact a vast abstraction which swallows up a myriad of choices and demands which arise from individuals.

At the risk of repeating myself, let me say again that I understand perfectly well that a group of men making individual demands and conscious choices in a common context *can* be interpreted from a more general point of view as constituting a self-regulating mechanism. My point, again, is that such an interpretation costs too much! We shall, a bit later in our discussion, be describing some of the findings of recent students of animal behavior. That branch of biology which is called ethology, the study of animal behavior, has made it perfectly clear that a good many animal and insect groups do indeed respond to "inputs" as a self-regulating unit through a mechanism of purely instinctive reactions in the various individuals. There is, for example, in Africa a kind of insect colony which can and does form itself into what appears to be a multicolored flower, thus disguising itself in the eyes of predators, with each individual according to its color taking just its proper place. Now here, with very little empirical distortion, is a genuine *system* of behavior—a real self-regulating mechanism of a social type. But of course, in this case the path of each individual is instinctively determined and not consciously chosen or even subconsciously chosen. It would, it seems to me, make a good deal more sense for ethologists, given premises comparable to those of social scientists, to attempt to devise an abstract theory of the system of animal behavior than it does for social scientists to construct an abstract theory of human behavior. Interestingly enough, however, the ethologists are not very much inclined to do this—for reasons which I shall try to make clear later on.

It will be helpful at this point to look again at the parallel between politics and science as a human institution. Easton, you will recall, attempts to abstract out of the whole complex of human behavior something which he calls the *political* system. He suggests that there is also the *economic* system and the *religious* system and both of these have boundaries which separate them from each other and from the political system. It is clear enough that given a sufficient amount of imagination and ingenuity we could in a parallel way talk about the *scientific* system. We could, if we really wanted to badly enough, define science and construct a theory of it as a sociological system. It would be possible to describe especially the contemporary scientific community as being characterized by authority structures, access points, feedback loops, and so forth. If we forced ourselves to be absolutely general, we could, following the example of our political science predecessors, look at primitive societies and see that, while the boundaries are fuzzy, the characteristic scientific functions are being *in some sense* performed by chiefs, medicine men, and Eskimo headmen. Notice, also, that we could, if we really wanted to, do the same thing with art or religion or literature. All of this is to say that the systems idea makes a pretty good metaphysic, a pretty good linguistic recipe, if one wants to put it to that use.

What we might call the systems theory of science is, of course, not altogether implausible, particularly in connection with the contemporary scientific community. Do we, however, really want to accept the notion of science as a self-regulating mechanism? Do we really want to swallow up the great creative acts of intellectual reorganization of Galileo, Copernicus, Newton, Lavoisier, and Einstein in some automatic abstraction? Do we really want to accept the notion that the invention of the rules of scientific inquiry was just a readjustment of the authoritarian system of knowledge of the Middle Ages? To accept such a general theory would be profoundly *nonempirical*; it would make something appear automatic which was in fact not automatic at all.

Let us think back for a moment over our earlier discussion of developments in the philosophy and history of science. Early in these two disciplines—say thirty or forty years ago—we find two quite separate enterprises. On the one hand there is the history of science, as described by Kuhn, the straightforward chronicling of particular scientific events. On the other we have the early philosophy of science, described by Toulmin and Hanson as an attempt to fix the pure, logical, abstract picture of the nature of scientific inquiry and scientific knowledge. Although this is to be sure still a matter of some controversy, in the eyes of Kuhn, Toulmin, Hanson, and others neither of these approaches is adequate to an account of what science is really all about. Science is neither a simple accumulation of accurate information nor the workings of a pure and perfect intellectual answer machine.

Although there probably never existed an absolutely pure case of either the history or the philosophy of science as just described, there can be little doubt that two quite different thought models or ideals are involved. The history of science so understood is a species of traditional narrative history. The abstract, pure picture of scientific inquiry is an instance of what might be called the Platonizing tendency always present in greater or lesser strength in philosophy. When Toulmin, Kuhn, Hanson, and the rest seek to provide a more accurate, more illuminating account of science, what in effect they do is to run history and philosophy together. They are content neither with the "tic, tic, tic of the clock" chronology which characterizes history nor with the ideal form picture of philosophy. The account of science which focuses on the successive invention of theories and paradigms of explanation is more than a history of science, it is a developmental, explanatory theory of science. It is intended to tell us not only what happened but why and how it happened.

The crucial factor in this theory of science is the treatment of time. It is neither a clothesline on which events are hung one after another nor is it abstracted out of existence by an ideal form. Time is neither ignored nor is it simply handled or recognized; it is *taken seriously*. This is the lesson that we have been struggling to learn in these pages—that genuine understanding involves taking time seriously as a fundamental aspect of reality. This involves more than constructing an essentially three-dimensional picture of a mechanism that *allows for* change over time. Historians or philosophers of science, as we have observed, would be unlikely to fall into this trap anyway, because their subject matter is rather obviously inappropriate to such treatment. Taking time seriously demands that time not be regarded as a kind of secondary factor that disturbs essentially static parts, but that it be understood as the primary factor which gives the apparently static components their meaning.

What I am getting at is rather like the relationship between a hydro-electric dam and the river on which it is built. One could look primarily at the dam and its component parts, noting secondarily that the flow of the river is subject to change and that the dam is built to accommodate a certain range of changes. Or one can see the mere existence of the river as primary and the dam and its various parts as only meaningful and significant relative to the river. What, after all, would be the significance of a dam built where there was no river?

Taking the time dimension really seriously is in fact very difficult because we are fortunately or unfortunately three-dimensional creatures or, more properly perhaps, we are *perceptually* three-dimensional creatures. In the deepest sense we can probably only get the "feel" of the time dimension by use of images like the river-dam just described or by employing what the philosophers like to call the *gedanken* or thought experiment.

Suppose, by way of illustration, that we try a *gedanken* experiment or two with a two-dimensional man. Of course, since there are no two-dimensional men, our experiment will have to be an "experiment in thought," but it has the value of making clear the sort of perceptual differences that result from dimensional differences. Consider, then, a man living and perceiving entirely within a Euclidean plane, in short, a two-dimensional man. From a three-dimensional point of view the plane is moving, passing through various three-dimensional objects. Suddenly a point appears to our two-dimensional man. It steadily expands, forming ever larger opaque circles. Finally the circles begin to contract, steadily, down to the original point which then disappears. Our two-dimensional man has seen a sphere. In order, however, for him to know that it was a sphere he would have to learn to take the third dimension seriously.

Suppose that he sees a point moving with regular motion in a circle. What he sees is in fact (or at any rate in three dimensions) a cylindrical spiral which is not moving at all, but he sees it as a point revolving around a center.

Strictly speaking, of course, it is impossible to draw a picture from a four-dimensional point of view. The addition of a little imagination can, however, help to convey the idea. Consider the drawing of motion through time of the sun and the earth on the following page. Notice now that the phrase which I just used, "motion through time," is from a four-dimensional point of view not really correct. To a being capable of perceiving the earth and the sun in full four dimensions the earth would not be in motion, but would appear—to use a three-dimensional image—as a fixed spiral. Likewise, any event or object—there would be no difference between events and objects, an event would be a "short" object and an object a "long" event— on the earth would, because of the earth's rotation on its axis, be a sort of spiral within a spiral. Furthermore, a complex of objects and events, such as a society, would be a set of irregulor spirals in a spiral in a spiral. A society would perhaps be something like a twisted telephone cable in which the wires were not regular.

Of course, we are dealing here not with precise descriptions of societies but with matters of perspective. And, once again, perspective shows itself to be all-important. By the phrase "taking time seriously" I mean to refer to the difference between seeing a society as a length of twisted telephone cable and seeing it through some generalized pattern stretched and loosened to fit each of an infinite number of cross-sectional slices cut from all of the twisted telephone cables.

A considerable variety of implications, some of them of substantial significance, follows from this change of perspective. The rest of this book will, in a certain sense, be nothing more than an exploration of these implications. For the moment, however, let me point out only two of the more basic and general ones. It is not scientific, but scientifically naïve, to

attempt to understand politics on the model of nineteenth-century physics. This is true not only because the fit between the model and the phenomena in question is not very close. It is true also because the static model systematically excludes the most important scientifically demonstrated fact about man. For man is, above all, a biological organism and he is what he is because of and through the process of biological and cultural evolution. Our putative science is and can only be a linguistic recipe unless this fact is recognized as primary. Thus, if we are to compare societies for the purpose of truly understanding them, we must compare twisted lengths of telephone cable and not mere cross-sectional slices. We must recognize that universal generalizations are inevitably vacuous, but we must at the same time understand that descriptive generalizations are the way to reveal patterns of political activity that take place along an evolutionary timescale. Stages or phases of political evolution must be understood as primary and the structures and functions of political organization must be seen as relative to the stage or phase of political development.

Making an argument for an evolutionary perspective on human affairs will inevitably invite a barrage of hostile criticism. The mere mention of the word evolution in connection with human social behavior automatically raises the specter of Hegel, Marx, Spengler, Herbert Spencer, and the Social Darwinists, and the whole style of nineteenth-century speculative social thought which we have all learned to disparage on some ground or another. There is, it seems to me, only one way to deal with such criticism and that is to face it squarely and inquire into its validity.

Perhaps the most general and most powerful line of criticism directed against this style of thought is what might be called the Critique of Historicism. In recent days this critical stance has been most often associated with the name of Karl Popper, but it was rather more profoundly (if somewhat differently) put earlier by Kierkegaard and Nietzsche. Understanding human affairs in a developmental, historical context can and has led to a tendency to ascribe causal influences to history itself, to a picture of the world dominated by inexorable laws of history. It is this view, most often attributed to Hegel and Marx, that is called Historicism and it is this picture against which the critics react. Let me note in passing that, so far as I can see, the Critique of Historicism is more precisely directed against the image of Hegel and Marx than against what Hegel and Marx actually said. The adjudication of this matter would, however, take us far afield, so let us suppose that we are dealing with Historicism in the sense of the inexorable laws of history.

The attack is essentially two-pronged. From a logical point of view it is argued that the Historicist creates a metaphysic built upon statements which are vague, meaningless, and useless. Thus, it is contended that statements like "The world is an all-encompassing spirit" and "All developments in human affairs are the result of more fundamental changes in economic relationships" are metaphysical, devoid of meaning, because no facts could possibly show them to be false. We are speaking again of "linguistic recipes" and of what might be called the "God-works-in-mysterious-ways syndrome." The ethical or, one might want to say, ontological criticism is that Historicism makes life meaningless in the sense that individual existence and choice are rendered meaningless by being swallowed up in the march of history.

These criticisms are quite sound and we must, I think, grant them. We can also assent to the objections raised against Spencer and the Social Darwinists for translating an evolutionary perspective into an advocacy of "survival-of-the-fittest" laissez-faire capitalism. Having said this as plainly as I know how, I can only hope that no one will choose to describe these pages as a twentieth-century revival of Social Darwinism, although I fully expect that someone will.

Having granted the validity of these objections, we must now ask what exactly have we agreed to? Nietzsche once wrote an essay entitled "The Use and Abuse of History" in which he was sharply critical of Historicism somewhat along the lines that we have just described, but in which he maintained that the historical frame of reference was crucial to understanding. Similarly, in accepting the criticisms of Historicism we give up the idea of History as a metaphysic, as what I have elsewhere called a "deductive absolutism," * but we demand its retention as the primary perspective from which we look at the world.

The point here can perhaps be most sharply made by saying a word or two about that perennial argument concerning whether Karl Marx is to be understood as a social scientist or as a philosopher-metaphysician. It is easy enough—I know this perfectly well, having often made this argument myself—to dismiss Marx as a mere metaphysician. In the simplest way the application of positivist tools will show that Marx's major premises concerning the all-controlling power of economic relations or the inevitability of certain historical developments are impossible to falsify in the eyes of a committed Marxist because he can always argue that the facts adduced in support of the objection are mere superstructure, but froth upon the inevitable wave of history. Thus, the positivist accutely argues if no conceivable evidence could be relevant to the refutation of Marx's premises, the premises cannot be empirical propositions, and Marxism, therefore, cannot be science.

It is moreover possible to go beyond this merely logical argument into the intellectual history and actual biography relevant to the question.** There is good evidence to show that Marx learned his historical, explanatory theory from philosophers Hegel and Feuerbach and that Marxism as originally conceived was a philosophical perspective in the tradition of German idealism. Furthermore, it can be demonstrated with a high degree of plausibility that the gathering of vast quantities of economic data later in Marx's life was done under the guidance of his inverted Hegelianism.

Saying all this, however, by no means demands that any standing for Marx as a social scientist must be denied. The inclination for the "tough-minded" social scientist to put Marx aside as a speculator, a mere phi-

---

* *The Logic of Democracy* (New York: Holt, Rinehart and Winston, 1962).
** See, for example, Robert C. Tucker, *Philosophy and Myth in Karl Marx* (London: Cambridge University Press, 1964).

losopher, derives partly from the influence of our old friend the hyper-empirical prediction-generalization model of science. Marx was clearly partly a philosopher and is suspect on that count alone, but in addition his generalizations and predictions do not hold up. There is, I think, no need to do more than mention the damage done to Marx's standing as a scientist by Marxism as an ideology.

The problem is of the classic "baby and bath water" type. However much Marx may have been a metaphysician, propagandist, or materialist theologian, neither these arguments individually nor all of them together are relevant to the validity of his historical, developmental perspective. We do ourselves a great disservice if in throwing out the bath water we throw out the baby as well.

Without, I think, making myself into a Marxist in any way, I assert that Marx was correct in his basic perspective. He took time seriously, even though his understanding of the significance of time was necessarily much less elaborate than ours can be in the middle of the twentieth century. The question of the time dimension in the study of human affairs really is, as Nietzsche described it decades ago, one of use and abuse. Because the evolutionary perspective has been abused and is always in danger of abuse does not mean that it has no use. I shall in subsequent chapters try to present what seems to me to be a useful application of the evolutionary, developmental perspective to political understanding. Let me now conclude this chapter with some remarks about the present state of affairs in political studies relevant to the point at issue.

We observed in a foregoing chapter that many natural scientists have found themselves forced to give up a Newtonian, mechanical paradigm of scientific explanation in favor of a developmental, historical one. What forced this change, we noted at the same time was an increased quantity and quality of information—in short, the nature of the subject matter itself. But we also observed—and it is perhaps this fact which is most instructive—that accepted paradigms of explanation die very hard indeed. Kuhn in describing the structure of scientific revolutions goes so far as to suggest that when a change of paradigm is proposed, what allows it to finally win the day (or the century) is not the conversion of its conservative opponents but their eventual death.

What seems to happen is something like this: Men are confronted with the extraordinary complex of factors and variables that make up their environment. Some man or group of men because of unusual mental ability or particularly fortuitous circumstances or some combination of both manage to clamp onto a set of environmental factors and organize them in a way meaningful for human problems. This "clamping on" may take place in any area of human activity, whether we call it art, philosophy, technology, religion, or science. When new facts, new problems, or examples of the "clamping on" of other men intrude, there seems to be an extraordinarily powerful tendency to ignore these disturbances or, if they cannot be ignored, to stretch, bend, and wiggle the original conception in order to capture the new facts or solve the new problems without giving up the old style of thought. When a new style of thought *is* accepted, interestingly enough and as Kuhn implies, it seems very often to be a biological matter. It requires a change of generation or a radical change of environment as when a scientist by some circumstance or other moves from one field of inquiry to a new one.

I suggest that we ought to attend closely to this picture, because—even if in a somewhat vague way—we may very well be talking about the mechanism of cultural evolution. Notice that in this account of the development of science—I would suggest that such a pattern applies to much more than science and perhaps to all of human life—we are taking time seriously and that this account of science itself is in the historical, developmental, evolutionary mold. If this is the kind of thing which that peculiarly human activity science is, then how foolish is our social science that does not give the central place to what human beings actually do. For men create scientific theories, works of art, technological inventions, monetary systems, constitutions, and electoral systems all for the purpose of solving their problems as they understand them. The social system as self-regulating mechanism cannot but be false as a general theory in the face of this. Its credibility, ironically enough, rests on a conception of science which is also false, but whether true or false this conception could not in any case be explained by a theory of the social system as self-regulating mechanism.

I said at the beginning of our inquiry into the nature of science that we had two objectives. First of all, we needed instruction on how we should proceed if we were to be scientific. And secondly, in discovering the nature of science we would also be discovering something important about the object of our inquiry, man himself. We have, I hope, by this time been able to cut through at least one layer of fog and are able to see that science makes no sense unless we see it as human creative activity developing through time and that man makes no sense unless we see him operating through time with his environment and with other men. More must be said and we have several more pages in which to try to say it, but let me now, as promised, say something about contemporary political science.

If it is the case that natural science is marked by strong conservative tendencies, by a powerful disinclination to give up old and well-worn habits of thought, then this same feature ought to be noticeable in political science. One of my former colleagues, who happened to be a student of Latin American politics, once defended the study of the politics of developing areas to me by saying, "It will save us from behaviorists." His comment is, I think, profound—more profound perhaps than he knew when he said it.

American political science until 1950 or so directed its attention almost exclusively to the politics of what we would now call modern societies. By far the greatest attention was given to the United States itself, and comparative politics or the politics of foreign governments meant almost entirely the politics of the modern states of Europe. Exclusive attention to modern societies meant that time could largely be ignored. Time was the business of American and European historians; the dynamics of the various political orders was the province of political scientists. Getting scientific about this subject could reasonably mean looking at the workings of these societies—the interaction of various groups, the behavior of leaders, and the voting behavior of followers. Thus, we found ourselves hearing about group theory, role theory, voting behavior theory, and so forth. And all of this—the so-called behavioral mood in political science—was constructed upon the foundation of the prediction-generalization model of science.

One fact we need to be quite clear about. The behavioral mood, however much and however often it may be described as revolutionary, is in truth not new-fashioned but quite old-fashioned. Its standard of explanation dates at least from Newton and in some ways from Plato. But even in its own territory, that of social science, "scientific" psychology and sociology are at least fifty years old and the advocacy of scientific method in politics heard in the 1960s does not differ markedly from the pronouncements of Merriam and Lasswell in the 1920s and 1930s. The behavioral mood's most important philosophical support, Logical Positivism, has been dead in the halls of philosophy for nearly thirty years.

After 1950 the subject matter of political science was forcibly expanded to include the politics of the new nations of Africa, Asia, and Latin America. Some political scientists obliged to deal with these areas found themselves suddenly doing history, even if history embellished with the technical terminology of political science. Some others, demanding of themselves a greater measure of scientific respectability, reached out to sociology and anthropology for what were presumed to be new methods and concepts. What they found *was* new to political science, but fairly old to sociology and anthropology and quite antiquated prediction-generalization stuff to science itself. Progress was surely made as a consequence of this borrowing, but exposure to the full sweep of the political subject matter causes the old prediction-generalization style models to virtually pop at the seams.

My point could not be better made than in the words of the best known borrower from sociology and anthropology, Professor Gabriel Almond of Stanford University. Almond in his book of 1966 with G. Bingham Powell, Jr. refers to the pressure of the subject matter on the old concepts in the following way:

> Our earlier formulation was suitable mainly for the analysis of political systems in a given cross section of time. It did not permit us to explore developmental patterns, to explain how political systems change and why they change. This static cross-sectional bias of the earlier formulations of the functional approach and, indeed, of much of the grand tradition of political theory and of contemporary political science research, raises some intriguing questions.
>
> Political science as an empirical discipline has tended to concern itself with the problems of power and process, with the *who* and *how* of politics—who makes decisions and how they are made. The *what* of politics, the content and direction of public policy, has generally been treated in terms of what political systems ought or ought not to do, or has been inferred from structure and process. . . .
>
> We need to take a major analytical step if we are to build political development more explicitly into our approach to the study of political systems. We need to look at political systems as whole entities shaping and being shaped by their environments.*

The sentiments expressed here are unexceptionable. Almond's theoretical formulation of 1960 in his well-known introduction to *The Politics of Developing Areas* was, as he here indicates, a classic example of an attempt to fit a set of fixed categories to all political events whatsoever. In 1966 the pressure of the subject matter made him recognize that his earlier attempt was inadequate and that the notion of development through time must be built into the theory. So far I would not and cannot quarrel.

When, however, one examines what is actually done in the 1966 book, it is clear enough that the full force of the time dimension has not been realized. There is much of value in the Almond and Powell volume and we shall be making reference to some of it later on. It is clear enough, however, that the developmental categories, the stages or phases of political development, are tacked on to the end of an analysis which remains fundamentally a static construction on the prediction-generalization model. The strength of old ideas is indeed great. What Almond and Powell did in 1966 is clearly a step in the right direction, but I would

---

* Gabriel A. Almond and G. Bingham Powell, Jr., *Comparative Politics: A Developmental Approach* (Boston: Little, Brown and Company, 1966), pp. 13–14. Almond's "earlier formulation" referred to in the quotation is Gabriel A. Almond and James S. Coleman, eds., *The Politics of Developing Areas* (Princeton: Princeton University Press, 1960).

argue that a few steps are not enough. What must be faced squarely is the matter of logical priority.*

One thing I think is clear about the relationship between a static theory of society and an evolutionary one. It is logically impossible to hold a full blown version of both at the same time. Either one or the other must be logically prior—the structures and functions must be relative to stages of development or the stages of development must be relative to the structures and functions. All we have said is an argument for the priority of the evolutionary and we must now look more closely into the implications of that choice.

* See the June 1964 issue of the *American Sociological Review* for a number of articles bearing on evolution and social science. The question of logical priority is, I think, never clearly faced. Note especially Talcott Parsons' "Evolutionary Universals in Society" on p. 339. See also Herbert R. Barringer, George I. Blanksten, and Raymond W. Mack, ed., *Social Change in Developing Areas* (Cambridge, Mass.: Schenkman Publishing Company, 1965) and Robert A. Nisbet, *Social Change and History* (New York: Oxford University Press, 1969).

7
The
Phenomenon of Man

1

The great American philosopher Charles Sanders Peirce once described truth as follows: "The opinion which is fated to be ultimately agreed to by all who investigate, is what we mean by the truth, and the object represented in this opinion is the real." * Thinking hard on this suggestion will, I think, repay our effort. Peirce is justly known as the father of American pragmatism, but unjustly credited with the idea that the latest and best conclusion which has been arrived at by science *is* truth. The well-known pragmatic postulate that what works is true may have been said and thought by someone, but it was not Peirce.

Peirce was a mathematician and his mathematical mind thought of truth as the ideal limit of a process. It is in this context that his famous

* *The Philosophy of Peirce, Selected Writings,* edited by Justus Buchler (New York: Harcourt, Brace and Company, 1940), p. 38.

categorical of fallibilism, "Do not block the way of inquiry," must be understood. Peirce did not argue that there was no truth or that truth was simply relative to the individual, but that no one could ever be certain that he knew the whole truth and was thus never justified in eliminating the possibility that he might be wrong.

Reality, Peirce suggests following Kant, is *there* even if we can never be certain that we know it. Reality has a relationship to human thought even if, again, we can never be certain that the relationship is one-to-one. The notion of truth as that which investigators would ultimately agree on suggests that we are often able to tell when we are wrong. The progressive recognition of error leads thought toward truth. Truth is thus to be approximated by institutionalizing the recognition of error, and this procedure, of course, is what we have come to call science. Scientific *method* —and academic enterprise in general would be much better off if this were widely understood—is not so much a positive thing as a negative one. It consists largely of a variety of tests designed to recognize error.

Truth or reality plays a kind of governing role with respect to human thought and this becomes more and more true as our culture becomes more scientific. It would be, I think, a mistake to say that reality "pulls" on human thought. It is more accurate to say that reality bears a relationship to the thought process like the environment does to biological evolution. Thus, reality is an ideal limit to the process of investigation in rather the same way that the environment is an ideal limit on the process of biological adaptation. We may indeed be talking about more than an analogy; we may be describing two aspects of the same process.

2

In the preceding pages there has been a discussion of several areas of human investigation. We talked about natural science itself, about the history and philosophy of science, and finally about the study of politics. In all of these areas we noticed a variety of misconceptions and false starts, but we also noticed with increased quantity and quality of information a tendency toward taking time seriously as an element in understanding. Biology, chemistry, and physics have become more historical and developmental as information has increased. The philosophy of science has become historical; the history of science more developmental like the history of philosophy. And political inquiry has been pulled into historical, developmental studies in a new way. This tendency, I suggest, is not a mere intellectual passing fancy; it is the work of reality shining through the spectacles of scientific method. I shall argue later on that this tendency is also operative in such unlikely places as analytical and existential philosophy and that it ought to be more explicitly understood in political theory.

These, however, are but themes that, like those in a Beethoven symphony, run through our earlier passages. It is now time to play the melody itself with full orchestra. To this end I have deliberately employed for the present chapter the title of the remarkable book by Pierre Teilhard de Chardin. For it is "the phenomenon of man," the whole biological phenomenon of human life that constitutes the melody from which our lesser themes echo.

My contention here is not metaphysical or religious. It remains, I think, within the realm of science. In arguing, as we have done, for the relevance of time to understanding we are opening the door to a broad view of the nature of man in the universe. A view of great breadth necessarily brings one close to the metaphysical. The notion of relativity in Einstein, of the "divine machine" in Newton, or of the uncertainty principle in Heisenberg have a metaphysical sound even though their purport is essentially scientific.

We can do no better in clarifying this matter than to quote the words with which Père Teilhard describes his own effort. Teilhard de Chardin as both Jesuit Father and distinguished palaeontologist no doubt was uniquely positioned to present a comprehensive view of man and the universe, but his view, according at least to his own words, remains within the realm of science. He begins his preface to *The Phenomenon of Man* * in this way: "If this book is to be properly understood, it must be read not as a work on metaphysics, still less as a sort of theological essay, but purely and simply as a scientific treatise. The title itself indicates that. This book deals with man *solely* as a phenomenon; but it also deals with the *whole* phenomenon of man." **

It is interesting to note that much of the commentary on the book since its publication has tended to ignore this explicit statement. *The Phenomenon of Man* is very often treated as if it were just what the author says it is not, "a sort of theological essay." This is perhaps partly a consequence of the widely known fact that the Vatican prevented publication until Teilhard's death and of the considerable discussion which the book provoked in theological circles when it was published. While it is true, it seems to me, that Teilhard was Catholic enough to demand large answers to large questions, he was also scientist enough to use science skillfully and to know what he was doing. We will be misled then if we do not take his own description seriously:

---

* From pp. 29–30, in *The Phenomenon of Man* by Pierre Teilhard de Chardin. Copyright 1955 by Editions du Seuil; Copyright © 1959 in the English Translation by Wm. Collins Sons & Co. Ltd., London and Harper & Row, Publishers, Incorporated. By permission of Harper & Row, Publishers, Incorporated.
** *The Phenomenon of Man*, p. 29. (Italics in original.)

In the first place, [the book] deals with man *solely* as a phenomenon. The pages which follow do not attempt to give an explanation of the world, but only an introduction to such an explanation. Put quite simply, what I have tried to do is this; I have chosen man as the centre, and around him I have tried to establish a coherent order between antecedents and consequences. I have not tried to discover a system of ontological and causal relations between the elements of the universe, but only an experimental law of recurrence which would express their successive appearance in time. Beyond these first purely *scientific* reflections, there is obviously ample room for the most far-reaching speculations of the philosopher and the theologian. Of set purpose, I have at all times carefully avoided venturing into that field of the essence of being. At most I am confident that, on the plane of experience, I have identified with some accuracy the combined movement towards unity, and have marked places where philosophical and religious thinkers, in pursuing the matter further, would be entitled, for reasons of a higher order, to look for breaches of continuity.*

Teilhard thus assures us—and to my mind the rest of the book bears out these assurances—that he is engaging in that classic procedure of the theoretical scientist, what Peirce called abduction or retroduction.** What he sees, and this is what we must see as well, are the facts of the extraordinary antiquity of the universe, progressive biological evolution, and the recent development of man and his creations. He retroduces in the sense that *The Phenomenon of Man* is an attempt to throw a consistent theoretical, explanatory net over these apparently disparate facts. His interpretations of the facts may, of course, be disputed, but so may any scientific theory be disputed. What Teilhard does, taking what he calls "biological space-time" seriously, is to sketch out the broad context in which human activity is set. And this, I suggest, is of great value to our investigation because it gives a substantive perspective to human political activity. I have argued that politics must be seen in an evolutionary context and Teilhard helps us to see what that context is.

3

Discussing man in a broad evolutionary context does nonetheless *sound* metaphysical. Because it is undeniably the case that metaphysics is very much out of fashion in the contemporary world, we are obliged to meet the question of metaphysics head on in a way which would have been quite unnecessary a century ago.

Teilhard continues his description as follows:

* *The Phenomenon of Man,* p. 29. (Italics in original.)
** See N. R. Hanson, *Patterns of Discovery* (London: Cambridge University Press, 1959) and Chapter 7 of my *The Logic of Democracy.*

But this book also deals with the *whole* phenomenon of man. Without contradicting what I have just said (however much it may appear to do so) it is this aspect which might possibly make my suggestions look like a philosophy. During the last fifty years or so, the investigations of science have proved beyond all doubt that there is no fact which exists in pure isolation, but that every experience, however objective it may seem, inevitably becomes enveloped in a complex of assumptions as soon as the scientist attempts to explain it. But while this aura of subjective interpretation may remain imperceptible where the field of observation is limited, it is bound to become practically dominant as soon as the field of vision extends to the whole. Like the meridians as they approach the poles, science, philosophy and religion are bound to converge as they draw nearer to the whole. I say "converge" advisedly, but without merging, and without ceasing, to the very end, to assail the real from different angles and on different planes. Take any book about the universe written by one of the great modern scientists, such as Poincaré, Einstein, or Jeans, and you will see that it is impossible to attempt a general scientific interpretation of the universe without *giving the impression* of trying to explain it through and through. But look a little more closely, and you will see that this "hyperphysics" is still not a metaphysic.*

4

I wish to stress the distinction between what Teilhard here calls "hyperphysics" and metaphysics because a great deal of error and confusion can be and has been engendered by a blurring of the distinction. The early applications of evolutionary thinking to human affairs illustrate this point with considerable clarity. A metaphysic (typically at least, although I am well aware of the fact that there are exceptions to this description) posits a "principle of reality" to which all events and occurrences are ultimately reducible. This is the significance of utterances, properly called metaphysical, such as, "However it may appear, everything that occurs in the universe is an aspect of the Absolute Spirit." Thus, in an evolutionary context, "survival of the fittest" could become the principle to which everything, no matter how unlikely it appeared on the surface, could be finally reduced.

There is a crucial difference between suggesting that human affairs are properly to be understood as embedded in an evolutionary context and attempting to reduce human activity to some sort of universal evolutionary principle. I am here arguing the former and emphatically not the latter—and this, I think, is Teilhard's position as well. Notice the difference between arguing in a Spencer-like fashion that since "survival of the fittest" is the principle of the universe that survival of the fittest

* *The Phenomenon of Man,* pp. 29–30. (Italics in original.)

in a dog-eat-dog economic sense is and should be the rule of the market-place, and the statement of Albert Camus, "The important thing, there-fore, is not, as yet, to go to the root of things, but, the world being what it is, to know how to live in it." * I here anticipate a point that I shall treat more fully later. For the present I want only to call attention to Camus's phrase, "the world being what it is." We are entitled to ask, as an empirical question, How is it? Camus is talking about knowing "how to live" in a certain context and we are, with Teilhard, attempting to describe that context, the context of human life. This can be understood as an empirical question demanding an empirical answer.

5

Of course, the first thing that we must try to grasp is the time dimen-sion. I choose the word "grasp" carefully for real comprehension is no doubt too much to ask of beings for whom the American Civil War is fuzzy and ancient history. Some years ago when I studied zoology at Indiana University, Professor William Breneman concluded his course with a lecture designed to illustrate the time dimension of biological evolu-tion. He called the lecture "From Kalamazoo to You" and in it he compared the development of the universe to a journey from Kalamazoo, Michigan, to the large lecture room in the Chemistry Building on the campus in Bloomington, Indiana.

I don't remember the proportions as Professor Breneman presented them in any exact way, but I can approximate his idea. The distance from Kalamazoo to Bloomington is roughly three hundred miles and if one takes the age of the universe to be something of the order of six billion years, one can say that the earth was formed at about Elkhart, Indiana, two hundred fifty miles away or about five billion years ago. No one, of course, can be very precise about the time at which life began on earth, but it was probably somewhere around Indianapolis, fifty miles and a billion years away. A sort of threshold of proliferation in the forms of life seems to have been reached about five hundred million years ago—twenty-five miles away or just north of Martinsville.

The final part of the journey was pictured by a length of string stretch-ing from one of the windows of the lecture room to a nail (which rep-resented the present) in the top of the laboratory table at the front of the room. A ribbon attached to the string six inches from the table would indicate the birth of Christ, one ten to twelve inches from the table would

* Albert Camus, *The Rebel: An Essay on Man in Revolt* (New York: Alfred A. Knopf, 1956), p. 4.

be the beginning of recorded history. Although no one is quite sure, it is generally agreed that the appearance of *homo sapiens* could be no more than ten or fifteen feet from the nail in the table top.*

Thus, it is clear enough that when we talk about politics from the time of the Greek city-state we are dealing with the last six inches of a three hundred mile journey. How curious it is that anyone should try to make universal generalizations using modern politics as a point of departure when modern politics has been with us in one small part of the world for about the last half inch of a three hundred mile journey—or to make the point a bit more justly but with equal power, the last half inch of a journey of fifteen feet, the time of *homo sapiens*. It is a confusion of the profoundest kind, but one which pervades social science, that modern, civilized Western man is the norm and that primitive or underdeveloped man is the deviant case. And it is a confusion that flows directly from the failure to take time seriously.

6

The question which we must face squarely if we wish to make scientific sense out of man and his activities is this: Does it make more sense to understand man as a biological phenomenon with all that that implies, or to try to fit human behavior to the prediction-generalization model of nineteenth-century physics?

When the question is put this way, the proper answer is, I trust, so obvious that it requires no discussion. However obvious the answer may appear when the question is so put, the fact is that modern social science has opted for the model of nineteenth-century physics. It would be, however, quite unfair to suggest that social science, at some time in the past, was presented with a clear choice like the one our question posits and stubbornly chose the model of physics, thus ignoring both the evidence and good sense. The choice was not clearly posed for a variety of historical reasons, some of which we have discussed earlier. At the time that the issue was decided the biological perspective was in an early and oversimple state and was often mixed up with some sort of historical metaphysics. The model of physics seemed at the time marvelously successful, pure, simple, and choosing it appeared honest and even heroic. The diffusion of ideas being what it is, it is not uncommon to encounter now— fifty years or so later—a masters candidate making this same choice in the name of honesty and feeling not a little heroic.

---

* 1,000 years would equal about three inches, 10,000 years about 2.6 feet.

7

The distinguished British biologist C. H. Waddington in discussing the biological role of human ethical systems finds himself confronting philosophers who are, like social scientists, under the spell of physics. He suggests that:

> The disagreement or even distaste and scorn which many modern philosophers evince towards theories such as I am putting forward here probably have their origin in rather deep-lying disagreements about what constitutes a convincing argument. Philosophical thinkers have, in the last few decades, been profoundly influenced by many advances in modern science. The advances which have made most impression on them have been those in the physical sciences. Open any book of the present day dealing with epistemology or the general problems of philosophy, and you will find a discussion of pointer readings, the theory of relativity, the quantum theory, the indeterminacy principle, and so on. These are undoubtedly exceedingly important matters, but one would have thought them somewhat remote from the general activities of human beings, except in the very special field of the quantitative analysis of the behavior of material bodies. Man is after all, a biological entity. It is only in his most generalized characteristic, which he shares with sticks and stones, that he is part of the subject matter of physics or chemistry. In his full being—or at least if we do not wish to beg the question, over a much wider range of his being—he falls within the province of biology.*

8

We are, as Waddington quite rightly suggests, dealing with "deep-lying disagreements about what constitutes a convincing argument" or what we have earlier called paradigms of explanation. One of the reasons why the issue between the developmental biological perspective and the prediction-generalization model has not been clearly focused until recently is that biology itself has only recently passed out of its own prediction-generalization phase. As Waddington himself describes it:

> A few decades ago the growing point of biological thought was the analysis of the operations of the living machine. The most advanced biology dealt with problems of metabolism, of the intake of oxygen, foodstuffs, etc. and

* From *The Ethical Animal* by C. H. Waddington. Copyright © 1960 by George Allen and Unwin Ltd. Reprinted by permission of Atheneum Publishers. p. 72.

the changes they undergo in the body. The fundamental problem of biology was seen as the understanding of the nature of enzyme action. More recently we have seen an increasing importance attached to questions concerning the mechanisms by which the functioning apparatus becomes gradually transformed as the individual develops from the fertilized egg onwards. This movement of thought, which had its origins in the work of such men as His, Roux, Driesch and Spemann, eventually and inevitably became linked with concepts derived from genetics. Its full depth and profundity then became apparent. The dominant position it now holds within the technical field of biology may be recognized in the fact that almost any biologist nowadays would admit that the crucial problem for theoretical biology is an understanding of the way in which genes control the characters of the organisms which develop from newly fertilized zygotes.*

Waddington's remarks here help us in two ways. First, as I have just indicated, this aspect of the recent history of biology helps us to see why the issue between the paradigms of explanation has been and in some ways still is a cloudy one. But in a more important way Waddington's comments speak to the question of logical priority and the "pull" of reality which we discussed a few moments ago. It is not that the problems of metabolism are not important or that they cannot be handled with some success on an input-output, prediction-generalization model. The point is that metabolism and enzyme action are, so to speak, "secondary" phenomena which could not exist at all, let alone in the particular way that they do, apart from the kind of animal that evolved and developed in the way that it has. The evolutionary, developmental questions are, thus, the more fundamental ones. They are logically as well as empirically prior questions. And this is so for no other reason than that animals and their functions are the products of an evolutionary process over a vast period of time and not the full-blown creations of some cataclysmic act a few thousand years ago.

9

In very much the same way, the attempt to make universal generalizations in human affairs derives from that rather static view of the world which was characteristic of biblical times and has biblical origins and which, of course, continued into the world of Newtonian physics. If it were indeed the case, as Newton and others of his time thought and as has been assumed until very recent times, that the world was simply created full-blown some six thousand years ago, then it would follow that the way to understand human affairs would be to look for regularities,

* *The Ethical Animal*, pp. 73–74.

for those universal principles that apply to all human life. Under those circumstances it would make sense to attempt to look at modern man first and primitive man second or primitive man first and modern man second. It really would not make very much difference because man would be man, and the principle would apply in any case. When one, on the contrary, looks at the problem from the perspective which evolutionary thought supplies us, it becomes clear that the important feature that we should try to grasp is not some set of universal laws that apply everywhere and for all time, but to try to see the way in which man has *developed* his peculiarly modern characteristics. Only in that kind of context can we begin to understand what modern politics is really all about. The evidence for this is the plain and simple fact that the best and most informative knowledge that we have about politics is historical knowledge—even if we sometimes present historical knowledge under the sociological category of "political culture."

The argument made here is clearly one which favors the historical, evolutionary point of view. I have tried to show the sense in which this perspective was rejected by twentieth-century social science as a consequence of certain historical circumstances, namely, the fact that the evolutionary perspective as it first developed in the nineteenth century was intertwined with a number of metaphysical assumptions and thus did not meet standards of scientific precision. Nonetheless, in order to deal with this issue fairly we must face the question whether or not there are still good reasons for choosing the universal-generalization model for understanding human affairs even though, as I have contended, the evolutionary model is no longer involved with the historicist metaphysic. To do so, to make this choice on the side of universal-generalization, seems to me to accept in effect the doctrine of special creation—something which one would not suppose that social scientists would ordinarily be inclined to do. But I think it must be said that if one is to contend that human behavior is some special sort of thing which does not follow an evolutionary pattern as man's physiological and anatomical nature clearly does, then this is to say that human culture is something fundamentally different from what we would ordinarily call man's biological nature. The question is are there good reasons for making this assumption or does it *simply* flow from the historical reasons which we have discussed earlier?

10

What then is the best argument that one could give to support the universal-generalization notion? I think that one of two notions would have to be presented. Probably the two together are the operative support for the choice. The first is that human culture—man's products and man's behavior—is something quite different from animal behavior. And the second notion probably is the assumption which we have discussed before, namely, that science simply *is* the seeking after universal generalizations. Now I think that we have rejected the second proposition with sufficient force and sufficient argument. Therefore we are left with the first, that is, that human culture is something different from biological development and consequently ought to be understood according to quite different standards.

At a certain level such an argument is obviously quite a plausible one. There is clearly a great deal of difference between the way in which man developed relatively small canine teeth and the way in which he chooses his political leaders. At a more profound level, however, the simple acceptance of a distinction of this sort and the basing of our science upon it involves an assumption which when examined carefully I am not certain that many social scientists would be prepared to accept. This assumption is very clearly focused upon by Teilhard de Chardin. He puts the problem in the following way: Either, he says, we assume that the peculiarly human kinds of behavior which involve consciousness and self-consciousness came into the historical development of the universe *from the outside* in some special way at some particular time, or we must assume that somehow even these characteristics grew out of the development of the universe and were, in some sense of the term, implicit in it from the beginning. Thus, we must ask the question if human consciousness and all that is connected with it were not somehow implicit in the historical development of the universe from the beginning, then where and from what source and under what circumstances *were* they introduced into the universal time process?

Surely then everything that we know about science and the natural order of things forces us to accept human consciousness, self-consciousness, behavior, and culture as *natural* phenomena. The case for a contrary point of view would seem ultimately to rest necessarily upon a doctrine of special creation or something quite like it.

Saying this, however, does not really decide the issue before us. It would still be possible to accept human behavior as a natural phenomenon and yet to argue that its proper understanding would be in terms of the universal-generalization model. What seems to be clear, however, is that the acceptance of human behavior as a natural phenomenon demands that we take evolutionary considerations as logically prior. We are, thus, obliged to look actively for the connection between man's biological development and his intellectual and cultural development. The presumption is that the two most certainly *are* connected and only a preponderance of evidence could persuade us that they are not. The burden of proof would, thus, seem clearly to be on the advocate of the universal-generalization model.

And this, of course, is how a paradigm of understanding works; it determines what kinds of questions ought to be researched while another paradigm might not even raise the question.

What I ask you to notice is that the evolutionary-developmental paradigm calls direct and crucial attention to the question of the relationship between biological evolution and social-cultural evolution, whereas this is not a question at all for structural-functional analysis or systems theory (at least as presently employed by social scientists). By the same token, while the universal-generalization paradigm by its basic logic forces someone like Easton to engage in extraordinary feats of concept manipulation in order to achieve universal categories and definitions, the evolutionary-developmental paradigm makes no such demand and on the contrary leads one to expect the remarkably sensible conclusion that different kinds of political orders are in fact different.

11

In pursuing the question of social-cultural evolution and its connection with biological evolution we face a certain difficulty. The difficulty is not so much a conceptual one as it is an expositional one. In order to do full justice to the evidence which is relevant, I should have to insert into these pages what would amount to a lengthy textbook on biology. Since I cannot presume to do so, I must ask you to accept, or at least entertain, a much briefer and more general account pointed specifically at the particular problem at hand.

Studies of biological evolution are, of course, vast and complicated. As in any research area, there is disagreement among experts over matters of detail and sometimes of interpretation. At the broadest level, however, there seems to be general agreement that while the course of evolution is certainly not wholly determined, neither is it wholly random. Thus, while biologists would generally argue that the process of mutation at the genetic level is random, they would also accept the notion that a direction is discernible in the course of evolution as a whole and often in the evolution of particular aspects of life. To put the latter point briefly, there seems to be little doubt that one can speak sensibly of "higher" or "more advanced" forms of life and that these higher forms on the whole developed later in the evolutionary process. Thus, in the most obvious case, man, who by a great variety of purely biological measures is the most advanced form of life yet produced also appears quite late in the time scale.*

* See, for example, the discussion by Marston Bates, *Man in Nature* (Englewood Cliffs, N.J.: Prentice-Hall, 1964).

The enormous complexity of the process and the fantastic diversity of the products cannot be overstressed. We need, nonetheless, in order to focus our vision, to make some general descriptive statements about the process of evolution. Everyone knows and understands the general notion that nature throws up an enormous quantity of chance variations in living things and that those which are able to "make a living" in the environment tend to survive while those that find no place perish. What we must notice is that this process depends upon what can perhaps be described as an "information transfer system." The most perfectly adapted particular animal or plant would be utterly insignificant from an evolutionary point of view unless it were able to transfer its adaptation to offspring. Recent research by geneticists into the DNA molecule has opened the door to a much more profound understanding of the system of information transfer from parents to offspring. It is not accidental that biologists speak of "breaking the genetic code" because what they are dealing with is the way in which information is *encoded* in the gene.

12

We nonbiologists find it easy enough to understand that particular *physical* characteristics such as the opposable thumb or the cloven hoof are genetically passed on through the generations. We understand that the giraffe has a long neck because it allows him to survive. But, I think, what is not nearly as well understood is that, particularly in animals lower on the scale than man, *behavioral* characteristics are genetically communicated as well.

When we speak of behavioral characteristics being genetically determined, we raise the notion of "instinct" and the paradigms of understanding again clash and grind like incompatible gears. No one, I think, could argue with the proposition that twentieth-century social science taken as a whole and including psychology, anthropology, sociology, and political science tends to be pretty skeptical about the idea of instinct. Test your own educational experience with respect to this point. We are taught, quite simply, that to talk about instinct is to be unscientific. In the field of psychology the use of instinct as a concept was for a long time classed as what Osgood described earlier as "junk-shop psychology." Of course, what is meant when it is suggested that instinct is unscientific is that it is "un-universal-generalization-scientific." For deeply embedded in the fundamental logic of the stimulus-response, input-output, structural-functional model of social understanding is the notion that any item of behavior can only be understood in terms of (in the sense of "scientifically related to") an environmental stimulus or set of stimuli external to the organism. The empiricist element in the model demands exclusive attention to observables and the universal-generalization element demands correlation of observables with an eye to the building of general laws. Recall Osgood's remarks quoted in Chapter 4, Section 3. The concept of parsimony which comes not from psychology but from the theory of science creates the "empty-box," and the empty-box, even when partially filled with "scientifically" respectable theoretical constructs, leaves little room for instinct.

Let me digress a bit further from the question of genetically communicated behavior and make some remarks in passing about a topic which I have so far avoided. At this juncture in our discussion it is possible to see quite clearly, I think, the great oversimplification involved in the idea of a value-free social science. The notion is widespread that it is possible by adopting the neutral stance of the physicist with respect to society that general and morally neutral truths about human relations can be discovered. And yet the adoption of the methodological standards of nineteenth-century physics leads inexorably to the twentieth-century environmentalist morality propagated far and wide by students of society in the name of science. If the social scientist's conception of science leads him to see behavior as caused exclusively by environmental stimuli (even though he may at the same time be mouthing moral neutrality, then he cannot but see the causes of social problems (dysfunctions, if you will) in the environment and solutions in environmental adjustments. Is it or is it not the conclusion of contemporary social *science* that crime is caused by environment and, in particular, slum environment? And do not deceive yourself with that fatuous dodge about the wearing of two hats, one of the scientist and the other of the citizen. The adoption of a world-view, the physicist's included, inevitably involves the taking of a stance which has moral implications. I do not suggest a strict deductive relationship between world-view and moral judgment—ethical decisions *never* work this way—but I do suggest that paradigms of understanding define the context in which moral judgments are made and, thus, crucially affect their content.

13

I shall have more to say about this matter a bit later, but let us return now to the genetic system of information transfer. The fact is that nature is replete with examples not only of genetically determined physical characteristics but also of genetically determined behavior. The situation is clearest in animals lower on the evolutionary scale than man. Countless instances have been observed and recorded, but none is more striking than those reported by the French entomologist Jean Henri Fabre and Dutch ethologist Niko Tinbergen.*

The larva of the Capricorn Beetle, which, Fabre tells us, is as unimpressive in appearance as a "bit of intestine," burrows into the trunk of an oak tree when it is as slim as "a tiny bit of straw." For three years it burrows its way through the inner depths of the tree, gradually increasing in size to the thickness of a man's finger. By ingenious experiments Fabre has studied the details of this creature's behavior until it finally emerges as a fully developed beetle. Throughout this period, the creature eats its way through the oak, the wood passing through its body into the tunnel behind; and it is in contact with no other creature, even of its own kind. But what happens at the end of this period? Fabre was able to prove that the fully developed beetle could not, by its own efforts, escape from the tree when only three quarters of an inch of wood separated it from the outside. Concluding that the larva itself must in some way prepare for the exit of the fully developed beetle, Fabre then studied the preparations for the pupa stage. Though it has never ventured near the outside of the tree before, the larva now makes its way to the bark. Sometimes it burrows through the bark and leaves an opening, but more often it leaves a very thin film of the bark unbroken. Then it retreats some distance down its tunnel and hollows out, in the side of this, a large chamber. This chamber, it must be noted, is not a closely-fitting one, adapted to the *present* size of the larva, but is sufficiently large to accommodate the developed beetle and to give some room for the action of its legs. The larva then constructs a cover, or door, to the entrance of the chamber. A pile of woody refuse is left outside, and inside the larva prepares a concave cover of chalky white substance which consists of carbonate of lime and an organic cement and is disgorged from the stomach. When this door is completed, the larva proceeds to rasp the sides of the now sealed-off chamber, covering the floor with a soft down, or wood-wool, formed by the minute shreds of wood. This down is then applied to the walls, forming a continuous felt at least a millimeter thick. When this task is finished, the larva sheds its skin and becomes a pupa, and its position is always such that its head is pointing towards the door of the chamber. The position itself is important, since the mature beetle has a stiff horn structure and could not possibly turn around in the chamber if by any chance the larva had taken up the wrong or reverse position. When the beetle is ready to leave the tree, it has before it only the white concave cover which can easily be pushed aside, and a pile of loose refuse. The spacious passage before it leads it,

* See J. H. Fabre, *The Wonders of Instinct* (English translation of papers from *Souvenirs Entomologiques*) (London: Duckworth, 1928), and N. Tinbergen, *The Study of Instinct* (Oxford: Oxford University Press, 1951).

without chance of mistake, to the exit, where it has, at most, only the very flimsiest film of bark to gnaw through.

Here, performed by a creature which looks no more than a "bit of intestine," is an example of intricate behavior which appears to be pregnant with foresight for the later stages of its own development, but which cannot possibly be learned from others, since it has been in contact with none! How are we to account for such behavior?

A second example of such apparently purposive behavior can be seen in the reproductive behavior of a more complicated organism: the three spined stickleback.

From the moment of hatching until the manifestation of reproductive behavior, the stickleback has associated only with the young of its own age and with its father, and after becoming independent it has only seen individuals in neutral condition, either males or non-pregnant females. In the spring, the increase in the length of day brings the male stickleback into a condition of reproductive motivation. This drives it to migrate into shallow fresh water and to undertake a kind of random, searching, exploratory behavior. Within a certain territory a nest is constructed which is more or less tubular and built in a depression of the sandy bottom with bits and pieces of stems, roots, and leaves. The underside of the male's body now has a vivid red coloration. Any other male who enters his territory is driven off, with specific fighting movements. When a female, with swollen abdomen and an upward tilted posture is seen, the male swims towards her in a zig-zag fashion. Then he swims away from her in the direction of his nest. If the female follows him he swims down to his nest and points inside it with his nose, whereupon the female goes inside. The male then performs a quivering motion, prodding the abdomen of the female with his nose, and, as a result of this action, the female spawns. The male- fertilizes the eggs and thereafter takes care of them, performing periodical fanning motions with his fins which keep the eggs adequately supplied with oxygen.

This is a simple description of the stickleback's reproductive behavior. . . . Even this, however, is sufficient to indicate the apparently purposive nature of this behavior, which is common to all male sticklebacks, which is admirably adapted to the needs of the stickleback and its young, but which is *not learned*.

Innumerable instances of such complicated behavioral processes—which cannot possibly be learned—could be quoted.*

Thus, it is clear enough that not only does nature find selective advantage in particular physiological or anatomical characteristics, she also finds advantage in what the animals *do* with their physical beings. The beetle larva who was, so to speak, programmed to behave as described was able to survive while those not so programmed perished without reproducing.

* Ronald Fletcher, *Instinct in Man* (New York: Schocken Books, 1966), pp. 19–21.

14

Before proceeding to the next step in the argument, we need to pause for a moment to say something more about our theoretical perspective. We have already said that it is possible to discern a direction to evolutionary development, and indeed it is. We will be seriously misled, however, if we suppose this progress to be a simple, linear, step-by-step matter. Nature is not goal-directed or systematic in achieving progress. On the contrary, evolution, taken as a whole, is much better described as a trial-and-error process, and nature can be best understood as pragmatic. To speak rather metaphysically, as I think we must if we are to grasp the proper perspective, nature is like a bundle of inexhaustible energy (forget the Second Law of Thermodynamics for a moment) expanding into a vast myriad of passageways. Some of the passageways are dead ends into which energy is directed for a time and eventually given up. Extinct species are represented by this sort of dead end. Some dead ends seem, however, to be viable—at any rate so far as anyone can tell. Thus, certain species, generally relatively low on the evolutionary scale, continue unchanged for millions of years and show no signs of altering. Still other passageways seem to be open to what might be called "lateral" expansion. It is as if the passage divides at its end into a nearly infinite number of capillaries and the energy keeps on expanding laterally to fill them. Consider the Galápagos finches or the variety of deer or land snails, for example. Finally, certain of the passageways seem to be open-ended and energy continues to push into them. It is here that we find man with his new form of information transfer.*

What I have said here is not intended as a precise description, but as an image in terms of which we can orient ourselves. Descriptive generalizations about this process are necessary in order to get it into words, but no universal law can be posited without falling into a variety of Easton's dilemma. Nature is and has been in the process of doing what works—not necessarily what works *best* because this would imply a perfect *telos* which we cannot be certain of, but simply what works.

While we cannot legitimately use the term "*best*" in this connection, "*better*" seems acceptable provided that we do not test it against irrelevant human philosophical inventions. For nature as a matter of fact *has* experimented with the big brain and produced a creature who has become biologically dominant, encircling the globe with his artifacts and the ocean of thought which Teilhard de Chardin calls the noosphere. In a natural context the fact that this has happened is the only test of "better"—if it

---

* Cf. Waddington, *The Ethical Animal,* pp. 125–126 and *passim.*

had not worked "better" it would not have been done—and this is what is meant when it is said that a direction in evolution can be discerned. But we have no reason to suppose that man is the best possible product. On the contrary, nature's invention of the big brain was probably, as Freud once remarked, "an uneven and careless piece of work."

15

If, as we have suggested, evolution is a pragmatic process which works through a mechanism of information transmission, then we must suppose that if a better system of information transmission were stumbled upon nature would capitalize on it. We can now risk a general statement which will not mislead us providing always that it is properly qualified. Thus, we can say that *other things being equal* nature will find selective advantage in creatures which are flexible with respect to their environment. The more able a particular creature is to adapt to varying environments the more likely it is that he and his kind will survive. At the genetic level a similar principle would lead to the selection of a bisexual genetic arrangement over an asexual one just because the possibility of adaptive variation would be greater—and of course, this has happened. But we are speaking of something beyond the purely genetic and we are saying that *other things being equal* creatures provided with more flexible equipment with respect to environmental adaptation will survive over those that have less flexible equipment.

In order to understand what is being said here we must keep the energy-passageways image clearly in mind. As stated, this proposition probably explains pretty well why *homo sapiens* developed once the primate line got started, or more specifically for example, why Cro-Magnon man replaced Neanderthal man in the European forests. But it cannot be stated as a general law without the "other-things-being-equal" because other things are typically not equal. Consider these remarks by Waddington about the evolution of insects:

> Evolution from a primitive arthropod to a highly evolved insect such as a fly or bee has undoubtedly involved the real improvement of the arthropod type of organization, but this improvement has at the same time brought with it limitations which render indefinite further improvements impossible. For instance, insects have adopted and brought to a high degree of perfection the system of aerating their tissues by means of small tubes leading in from the exterior, through which oxygen passes by the process of diffusion. This system can be extremely satisfactory when the distance which the oxygen has to travel is small; but it would be almost impossible to evolve from it a respiratory system which could serve the needs of animals much bulkier than

the largest insects already are. Now, this limitation of the total volume of the body clearly limits the size, and therefore the possible complexity, of the nervous system. One cannot see any way in which an insect could evolve a brain as large as that of a human being. Although there is probably no precise relation between the size of the brain and the complexity of mental processes, they undoubtedly go roughly in parallel with one another. It seems almost certain that in adopting and perfecting their particular mode of sup-plying their tissues with oxygen, the insects have cut themselves off from the possibility of evolving a highly complex nervous apparatus. Similar consid-erations probably apply to all the major groups of the animal kingdom.*

Thus, the insect line taking as it did a particular passageway cannot be said to have developed in the direction of greater individual flexibility. While it is true enough that there has been a movement toward greater complexity of social relations—bee, ant, and termite societies are extraor-dinarily complex—the concomitant development of individuals has not been toward greater flexibility but on the contrary toward an increasingly rigid genetically determined program of behavior. The worker bee is not free to connive at becoming queen. Nature takes a variety of passageways and does what works for each passageway. A decision, so to speak, once taken determines the range of future decisions.

That most flexible piece of equipment, the big brain, therefore de-veloped only along a particular passageway. It developed in the strain of warm-blooded animals and this was probably necessary, as Julian Huxley suggests, in order to establish relative independence from the environment. A certain body size was probably required, so that, for example, aero-dynamics together with the amount of motor energy that could be utilized probably cut birds off at some point from further brain development. Presumably the upright stance, the terrestrial as opposed to the arboreal environment, and the opposable thumb all played an adaptive role. Some-thing is known about the details of this story to be sure, but much more information must be gathered before the account is complete. One thing is certain, however, and that is that man, with his peculiar brain *did* develop and that he was an offshoot of the warm-blooded, mammalian, primate line. Thus, if we are to find the connection between biological evolution and human cultural evolution it is to this warm-blooded, mammalian, primate passageway that we must look.

16

Several decades ago, toward the beginning of serious evolutionary studies, there was a great deal of dispute over the question of whether or not acquired characteristics could be inherited. Nearly everyone, save

* *The Ethical Animal,* pp. 134–135.

perhaps some unreconstructed Stalinist, is now persuaded of the fact that a characteristic acquired by an individual during its lifetime cannot be genetically passed on to its offspring. The demonstration of this fact has been a considerable scientific triumph. Biologists, however, are no less immune than anyone else to the inclination to state the implications of a truth a bit more broadly than it quite deserves.

Most biologists at the present day, in expounding evolutionary theory, seem content to leave it that the mechanism by which evolution has been brought about is composed of these two major factors: the genetic system with random mutation on the one hand and natural selection on the other. The evolutionary pressures exerted by these two factors are exhibited as being quite external to the nature of the organisms involved. The essential evolutionary pressure exerted by the genetic system is that of mutation, and mutation, it is explained, is a random process. Any explanation which might be offered for the nature of mutational changes would have to be found, it is asserted, in the chemical composition of the genes and not in the nature of the complete biological organism in which these genes are carried. Mutation thus appears as essentially an external force to which the organism passively submits. Again, natural selective pressures are usually thought of as arising simply from the external environment. When the climate changes, a new predator appears, or industrial fumes blacken the tree trunk on which the animal lives, the populations of organisms concerned cannot, it is usually implied, do anything but submit to these pressures and wait until the equally uncontrollable process of mutation throws up a new hereditary variant which enables them to meet the environment's challenge more successfully.*

No one can challenge the essential correctness of the general theory of the interaction between mutation and natural selection. As Waddington suggests, however, the conclusions can be overdrawn. A helpful way of making the point that I want to establish here is to notice the fact that it would be an enormous advantage to evolutionary success if it *were* possible for acquired characteristics to be passed on. One could reasonably speculate that if nature were to find a way to reap the advantages which would accrue from the passing on of acquired characteristics, she would surely take it. To put it less animistically, a species which developed some sort of mechanism for the transmission of adaptive advantages would be selectively favored over one which did not. If such a passageway were to be available, one would expect that it would be taken.

Waddington, speaking as the distinguished embryologist which he is, suggests that there are good reasons to loosen the rigid account of the mutation-natural selection process described above:

In my opinion, biology has already made all the discoveries of matters of principle which can be reached by this way of formulating the situation.

* *The Ethical Animal,* p. 88.

The time seems to have come when we need to take into account two further aspects of the evolutionary mechanism. In the first place, natural selective pressures impinge not on the hereditary factors themselves, but on the organisms as they develop from fertilized eggs to reproductive adults. It is only by a piece of shorthand, convenient for mathematical treatments, that indices of selective value are commonly attached to individual genes. In reality we need to bring into the picture not only the genetic system by which hereditary information is passed on from one generation to the next, but the "epigenetic system" by which the information contained in the fertilized egg is translated into the functioning structure of the reproducing individual. As soon as one begins to think about the development of the individuals in an evolving population, one realizes that each organism during its lifetime will respond in some manner to the environmental stresses to which it is submitted, and in a population there is almost certain to be some genetic variation in the intensity and character of these responses. *Natural selection will favor those individuals in which the responses are of the most adaptive value.*

Two consequences can be expected to follow, and have in fact been demonstrated experimentally. In the first place natural selection will build up genotypes which set going developmental mechanisms which easily respond to environmental stresses by the production of a well-organized modification which is of adaptive value. It will, as it were, build into the genotype a gun which is not only set on a hair trigger but which is aimed to hit the target when it goes off. In so far as such a developmental response becomes precisely delimited and easily initiated, it becomes more likely to be produced by unspecified changes in the chemical nature of the hereditary substance. Mutations, which we can think of as random when we are considering nucleoproteins in the chromosomes, will have effects on the phenotype of the organisms which are not necessarily random, but which will be modified by the types of instability which have been built into their epigenetic mechanisms by selection for response to environmental stresses.*

Waddington, thus, finds evidence for a sort of feedback relationship between epigenetic development and the genetic mechanism itself. Nature will and apparently has found ways of taking advantage of the life experience of individual organisms for evolutionary development. It is from this kind of perspective that we can begin to see what man is doing here. Nature has found a way in the warm-blooded, mammalian, primate passageway to increase the capacity to adapt by adding to instinct the capacity to learn, and finally not only the capacity to learn but the capacity to teach as well. While we certainly are not in a position to specify all of the precise details, the evidence is substantial that in the human passageway nature found a new method of information transfer, namely, that product of human learning and teaching that we call culture.

* *The Ethical Animal,* pp. 89–90. (Italics added.)

And through this system of information transfer evolution has continued, giving man the ability to fly, to stay beneath the sea, and to modify his environment in thousands, perhaps millions of ways. It is from this perspective that we must see the phenomenon of politics, remembering always that however important the utilization of natural resources may seem (as, for example, it did to Marx), the cultural system is basically a "sociogenetic" system, a system of information transmission.

8
The
Phenomenon of Politics

1

What meaning can we reasonably attach to the phrase, "Man is by nature a political animal"? If we are to talk about politics as a biological phenomenon, this is at least the sort of question which we shall have to face. The phrase, of course, belongs in the first instance—at least so far as we know—to Aristotle. There is, it seems to me, a great deal to learn from the careful examination of this Aristotelian premise. On the one hand there is the overriding question in what sense, if any, can Aristotle's assertion be said to be true? The way in which this question is answered sets the whole framework for what will count as political understanding. Thus, on the other hand what on the surface at least appears to be a separate question emerges as vitally important: What did Aristotle himself mean by the phrase and why has his meaning been so largely rejected in the modern world?

I am suggesting here that there appear to be two sorts of question.

One is presumably empirical: Is man a political animal or not? The other seems philosophical: How did Aristotle mean it and why has this meaning been rejected? My notion is that these questions are not so separate as they appear. To pull them together will, however, take some showing. I must therefore ask you to put yourself in my hands for a few pages while I plunge off into what may look like a digression. Hopefully, by the end of this chapter it will be clear that we are on the main track.

2

What then *did* Aristotle mean when he said, "Man is by nature a political animal"? The best point of departure in dealing with this question is to refer back to our earlier brief discussion of Aristotle in connection with the nature of paradigms of explanation. Recall Toulmin's treatment of cooking, ripening, and Aristotle quoted in Chapter 3, Section 3. There, you will remember, it was suggested that Aristotle's understanding of the nature of things was essentially biological. He saw the principle of nature as one of growth according to pattern. Thus, the process of cooking would have been understood by Aristotle in terms of the process of ripening and not the other way around. Understanding any particular thing or process was not for Aristotle a matter of breaking it up into its component parts and seeing how they fit together—an essentially mechanical mode of analysis—but on the contrary a matter of discovering the end toward which a thing or process was growing and seeing how it was so growing. Thus, it is correct to say that Aristotle's "root metaphor"—to use Pepper's term—was a biological one. It is important to see at this point that Aristotle's mode of analysis was biological not simply with respect to what we would now call "biological things," but in general, with respect to anything and everything. This—and this remark is a genuine digression—is what makes it so unsatisfactory to treat, say, Aristotle's *Politics* in isolation from his *Physics, Ethics, Poetics,* and indeed from his *Posterior Analytics.*

Aristotle's thinking, while certainly biological, was, however, by no means evolutionary. He saw—very empirically and in a very common-sense fashion—a sort of constant pattern of birth-to-death cycles. His judgment was that all particulars are to be understood as participating in a certain pattern of movement from potentiality to actuality. All little pigs grow into big pigs and all acorns grow into oak trees—or, more precisely, these cycles and countless others like them obtain if there is no external inhibitor. Aristotle was therefore able to conclude with the general statement: All things tend toward a certain end. Understanding a thing, thus, necessarily involved understanding the end toward which it was tending. This is obviously a very simple description of the Aristotelian doctrine, but I think it will suffice for present purposes.

Aristotle's question with respect to man was, therefore, what is the end for man? The whole of the *Nicomachean Ethics* is devoted to answering this question in full. The substance of the answer, however, is simply that the end for man is happiness and that happiness involves the contemplative life. Aristotle recognized that man shared many characteristics with other animals, but that what differentiated man's particular potentiality-to-actuality pattern was his capacity to develop the rational faculty. And Aristotle concluded in a way that all of modern psychology and anthropology can scarcely contradict that the development of reason required the development of speech which required interaction with other human beings. Thus, man was by nature the rational animal, the speaking animal, and the social (political) animal. Of course, Aristotle, when he used the term "political," meant something more precise than merely "social"; he referred to the polis and to the possible perfection of speech and reason in it. This matter is thoroughly discussed elsewhere and we need not be detained by it.

What is important for purposes of the present discussion is to see that the phrase, "Man is by nature a political animal," was for Aristotle embedded in a general view of the world which saw nature as consisting of an established set of patterns or cycles, and that each particular thing moved according to the particular cycle or class of things to which it belonged. Thus, again, acorns moved toward becoming oak trees and little pigs toward becoming big pigs, but an acorn could never become a big pig or a little pig an oak tree.

Aristotle, of course, had no real conception of the evolutionary reasons why this should be so. He simply saw that it *was* so in a very large number of cases and he extrapolated, so to speak, from these cases to all cases whatsoever and, thus, to a general principle of the nature of things. Another way to say essentially the same thing is that Aristotle began his analysis from an implicit and all-encompassing premise that nature *was* ordered and that man like all other things was a part of that order. Thus, knowledge and understanding were essentially matters of *discovery* rather than of invention. From this point of view, the world *is* ordered—there can be no doubt of this—and man's task if he wishes to solve problems is to find out what that order is and act in accordance with it. Notice that from this perspective there can be no question of a logical gulf between is and ought. "Oughts" are obviously derivable from knowledge of a thing which necessarily involves understanding what its end, purpose, and perfection is. Thus, the notion of "moral knowledge," which from a more recent point of view is a contradiction in terms, is an obvious and important kind of knowledge.

3

Before proceeding with a description of that "more recent point of view" just referred to, let me pause to put a question in your mind which I do not propose to try to answer immediately. Granted that Aristotle held this ordered view of nature and that his view is more or less typical—at least in logical type—of what we have come to call the classical mind, the interesting question is simply, "Why did he hold such a view?" I do not mean "why?" in a personal psychological sense, but rather something more like, "What was it in Aristotle's cultural circumstance and in the cultural circumstances of the countless others who agreed with him that made this way of looking at things seem correct?" When taken seriously, this is a very difficult and yet crucially important question. Let us allow this question to sit for a moment and move on to discuss the modern rejection of the Aristotelian view.

4

I suggested a few moments ago that Aristotle's biological perspective pervaded all of his work including his *Posterior Analytics,* in short, his very conception of what it was to be logical. Aristotle held that knowledge was obtainable by deductive demonstration from a general premise. Consider his classic syllogism:

All men are mortal
Socrates is a man
∴ Socrates is mortal

Examining this syllogism will get us to the heart of the difference between the classical view of things and the more modern one. Let us ask the question, Why, or in what sense, does the conclusion follow from the premises? and try to answer it from a modern point of view. I should perhaps interject here that I am speaking in very general terms when I use "classical" and "modern" and have no reason at present to go into the multitude of variations which a historian of philosophy might want to bring up. From a modern view, then, the conclusion can be said to follow from the premises because, and only because, it is *implicit in* the premises.

What is being suggested is that a syllogism of this sort depends for its validity on defining words in certain ways. Thus, it is argued that the conclusion provides no knowledge about the empirical world, it only reflects the fact that the premises have been set up in a certain way. A syllogism, therefore, is said to be a kind of tautology. Restated, it becomes, "If we define 'men' to include the characteristic 'mortal,' then Socrates, defined as being a man, is mortal." The principle from which all of this is derived—we touch again on our earlier discussion of the verification perspective—is the notion that knowledge of the empirical world can only be stated in propositions that can be tested against sense experience and found to fit or accurately describe that experience. Thus, "All men are mortal" cannot but be a definitional statement because we can show only that men who are dead are mortal and we cannot be certain about those now living or those not yet born. "All men are mortal" is not, therefore, an empirical statement, but a way of defining man.

Aristotle, of course, looked at this matter in quite a different way. Remember now that he starts from a perspective which has nature *ordered,* more precisely, ordered in terms of a constant set of potentiality-to-actuality cycles. Thus, when one observes a number of men and sees that they are mortal, it is possible to conclude not simply that these particular objects are mortal, but that one has seen the essential nature (potentiality-to-actuality cycle) of the category "man." "All men are mortal" therefore is not simply a definition, it is a description of the natural class of things called man. Remember that Aristotle invented the notion of biological classification by species and genus. Reflect also on Aristotle's well-known definition of a definition.

5

What I am suggesting by this discussion is that there is a fundamental difference in perspective between classical and modern and that recognizing this difference is of very great importance to our understanding. From the classical point of view man is very much *in* nature. He operates as a natural phenomenon in the midst of other natural phenomena. The instincts, the emotions, the wants, desires, and dislikes are real and personal. Knowledge about man in general and one's self in particular must always be practical knowledge in the sense of having genuine consequences. Knowledge can never be detached.

The modern mind, on the other hand, tends to put itself outside of nature looking in. The distinction between subject and object is clearly and sharply drawn. Historians of philosophy generally date the beginning of modern philosophy from René Descartes. What distinguishes Descartes from his predecessors? The radical separation of mind from body is the distinguishing feature. Hannah Arendt in an acute metaphor describes the rise of modern science as the discovery by Western man of the Archimedean point, the place to stand *outside* the world from which man can lift the world. John Locke in the fundamental statement of philosophical empiricism describes the mind as a *tabula rasa,* a blank slate detached from nature upon which experience writes. The mind-body dualism, the Archimedean point, and the *tabula rasa* all reflect the typically modern position of detachment from nature.

6

Let us now return to the proposition (or putative proposition if we want to be modern about it) "Man is by nature a political animal." From a classical point of view what is being described is one of the essential characteristics of the class of things called man. From the knower's position *within* nature ends and purposes were as real, certain, and genuine as anything else. Why, on the face of it and without the intervention of any theoretical criteria, should the fact that men are tending toward some end be any less real than the fact that men tend to have five fingers on each hand? From this position within nature it would be equally clear that things other than man have purposes. Do acorns tend to grow into oak trees or not? If you are inclined to dismiss all this as simple anthropomorphism, I must ask you to reflect on two things. First, whether you want to call it anthropomorphism or not, a great many people for a very long time did in fact think this way and, I suggest, it is important that this be recognized. Secondly, ask yourself the question: What kind of theoretical perspective is necessary for me sensibly to describe this kind of thinking as anthropomorphic?

Thus, man *feeling* (I choose this word carefully) himself inside nature would see man as literally *by nature* a political animal. As soon, however, as the knower takes up a detached position *outside* nature, the sense of the phrase must change radically. This perspective leads the knower to look at all features of nature including man as objects—"objects out there to be examined." From this perspective one must make a premise of doubting all premises and one must be limited to calling true only that which can be certainly seen to be the case. Thus, the suggestion that all men are political animals taken as an empirical statement becomes in principle uncertain, and the phrase "by nature" empty of meaning. From here, of course, in the realm of logical analysis the kind of logical tests described as modern a few moments ago come into play.

To anticipate what I shall try to say more fully later on, this change in perspective alters the whole fabric of understanding that goes into the contemplation of man's political activity. Politics becomes not as it formerly was a matter of discovering man's proper place in nature and acting in accordance with it, but a matter of describing objects and *inventing* a way to deal with them. It is, I suspect, no accident that the idea of the social contract which was but a minor theme in classical political thought becomes the dominant consideration in the wake of the acceptance of the new detached perspective of analysis.

Looking at nature, including human nature, from the outside makes politics a matter of construction, of invention. "By art is created the great Leviathan," Hobbes asserts, and not, as was formerly thought, by nature. The Hobbesian perspective is nowhere more clearly described than by Sheldon Wolin:

> "Science"—to use Hobbes's comprehensive term—had progressed so rapidly because scientists had been bold enough to break with traditional modes of thought and inquiry. They had refused to follow the path of building slowly on past achievements, of zealously preserving the main corpus and modifying only where necessary. The unprecedented development of "science" was pictured by Hobbes as an intellectual drama of creative destruction. Men had taken a radically new look at the universe, shedding their preconceptions and purging from their categories the vestiges of Greek teleology and Christian cosmology. By intellect alone, without appeal to superauthority and without relying upon non-rational and non-sensory faculties, man had created a rationally intelligible cosmos without mystery and occult qualities.
>
> Deeply impressed by the dramatic potentialities of this procedure, whereby man created intelligibility among the phenomena of nature, Hobbes then turned to convert it to the uses of philosophy, to make creative destruction the starting point for philosophical method. True philosophizing commenced with what Hobbes designated "privation"; that is, an imaginative act of destruction, a "feigning the world to be annihilated." . . .
>
> What was breathtaking about the enterprise was that it rested upon a conception of truth not as a faithful report of external "reality" but as an "arbitrary" construction of the human mind. . . . The crucial point, however, was that for Hobbes the "arbitrary" and the creative were synonymous. Hobbesian man emerged as the Great Artificer, the creator of science, mathematics, and philosophy, the architect of time and space, values, and truth itself.*

7

We now find ourselves in a very curious situation with respect to attaching meaning to "Man is by nature a political animal." If, on the one hand, we accept the classical position with all that it involves, the proposition appears to be simply true. From a modern, empiricist per-

---

* Sheldon S. Wolin, *Politics and Vision* (Boston: Little, Brown, 1960), pp. 245–246.

spective, however, the canons of logic direct us to regard the statement as either a meaningless piece of metaphysics or simply a sort of prescriptive definition. But, as the case of Hobbes illustrates, the modern perspective does not necessarily leave it at this, but may go on instead to show man as the political animal in a new sense, as the creator of political order.

8

There are, I think, two important things to be learned from this apparent paradox. The first we have already in a certain sense discussed at some length. Put in general terms it can be stated something like this: Human thought seems to work according to general organizing principles, *Gestalten* if you will, or paradigms of understanding. Here the example of the Neapolitan ice cream pie becomes relevant again. It all depends on how you cut into it! The process appears to work something like this: The mind settles on a certain aspect of a situation, comes to see it in a certain way, and then either ignores other aspects of the situation or forces them to appear consistent with the central focus. I am, I think, quite convinced that this is the way that the minds of philosophers, including natural philosophers (Kuhn's argument speaks to this point), work, although I am less confident about it as an absolutely general statement. Some psychologists would surely argue that the ordinary man quite readily holds inconsistent beliefs. But one can always ask inconsistent from whose point of view? Moreover, the psychologists surely recognize the stresses caused by what they call "cognitive dissonance," and the tendency to reduce or eliminate that dissonance.

However this may be, it seems clear enough that in the cases now before us, a focus on a *certain aspect* of the human situation has a sort of controlling effect on the *general* understanding of the human situation. Thus, if one settled on the notion of an all-encompassing order in nature as Aristotle did, then the very canons of logic itself, together with the understanding of the nature and limits of human knowledge, become subordinate to it. This, it seems to me, is why the jump from the empirical (in the modern sense) observation that "some men are x" to general statements (All men are x) about the category man seemed to bother Aristotle very little. It clearly did not bother him enough to make him reject his substantive argument.

Quite the reverse seems to be the case with someone like Hobbes and the many thinkers who follow in his wake. Here a stress on the epistemological and logical aspect of the human situation seems to result in a complete reevaluation of man's life in general and his political activity in particular. One does not start with community as a premise (given by nature as it was for Aristotle), but with the "state of nature" which by definition is the absence of community. And we are by no means merely speculating here, for, as Wolin points out, Hobbes is quite explicit about his epistemological and logical premises, calling for a creative destruction of the accumulated falsehoods of the classical and Christian tradition.

It is important to notice that while the political understanding of Aristotle and Hobbes differs sharply in relative emphasis, their views are not *utterly* different. They are both, after all, talking about politics. Aristotle stresses community as given by nature, but he does not ignore man's rational capacity to create by political choice. Hobbes in stressing the rational capacity for political choice directs his attention at choosing to create community. Thus, difference in theoretical perspective is very often a difference in emphasis. And, I think, this is probably especially so in political theory. Democratic theory, for example, nearly always involves some discussion of both majority rule and individual rights, but the difference in emphasis can be very great.*

9

So much for lesson number one. Lesson number two can be simply, indeed perhaps tritely, stated as follows: Politics almost certainly involves what might be called a "by nature" aspect *and* a rational, creative aspect. I suggest that this statement might be considered trite because surely any ordinary, intelligent observer can see that men "naturally" exhibit behavior which indicates group loyalty, personal interest, desire for power, deference to leaders, and so on; and that politics also involves creative choice with respect to public policy. It takes professional political scientists to confound the issue. For a great debate rages, or at any rate has been raging, over whether politics should be understood in terms of a set of general statements about the ways in which all people behave in group situations or in terms of the principles involved in making wise policy choices. In the name of intellectual purity there is a strong inclination for each side to try to drive the other out—on the one side into philosophy and law and on the other into sociology and psychology. There are, it is fair to say, a considerable number of would-be mediators to be found in political science, mediators who argue that both aspects should be included in the study of politics. Most of the time, however,

---

* I try to shed some light on this matter in the introduction to *Plato: Totalitarian or Democrat?* (Englewood Cliffs, N.J.: Prentice-Hall, 1963) and in Chapter 9 of *The Logic of Democracy* (New York: Holt, Rinehart and Winston, 1962), reprinted in Bishop and Hendel, *Basic Issues of American Democracy* (New York: Appleton-Century·Crofts, 1965), pp. 102–110.

the mediators restrict themselves to exhortation and never really face the crucial underlying philosophical question, which is, simply: How, under what kind of theoretical perspective, can the merger be accomplished? An empiricist perspective, as ordinarily understood, is, because of its philosophical focus, inevitably relativist with respect to the principles of policy choice. The professions of neutrality (as opposed to an avowed position of relativism) with respect to matters of value are, as we noted earlier, chimerical. Similarly, the value-oriented cannot with equanimity accept the implications of a thoroughgoing modern empiricism.

10

But still, even in the face of these theoretical puzzles, we somehow know that both a "by nature" aspect and a rational, creative aspect are part of political activity. Let us, therefore, ask our question again—What meaning can we reasonably attach to the phrase "Man is by nature a political animal"?—and try to give it a straightforward answer.

> Cities [says Claude Lévi-Strauss], have often been likened to symphonies and poems, and the comparison seems to me a perfectly natural one: They are, in fact, objects of the same kind. The city may even be rated higher, since it stands at the point where Nature and artifice meet. A city is a congregation of animals whose biological history is enclosed within its boundaries; and yet every conscious and rational act on the part of these creatures helps to shape the city's eventual character. By its form, as by the manner of its birth, the city has elements at once of biological procreation, organic evolution, and aesthetic creation. It is both natural object and a thing to be cultivated; individual and group; something lived and something dreamed; it is *the* human invention, *par excellence*.*

Lévi-Strauss, in here noting the conjunction of nature and artifice in human society, in a general way at least echoes the insights of the *Gemeinschaft und Gesellschaft* theorists of German sociology, the views of his countrymen Henri Bergson and Jacques Maritain,** as well as those of a considerable number of other observers. Indeed, one might argue—and I suspect that Maritain, for example, would be inclined to insist—that Aristotle himself saw the matter in this way. I would not want to argue that Aristotle was not in some ways correct, but I do want to insist

---

* Claude Lévi-Strauss, *Tristes Tropiques* (New York: Atheneum, 1961), p. 127. (Italics in original.)
** See, for example, Bergson's *Two Sources of Morality and Religion* (New York: Henry Holt, 1932) and Maritain's *Man and the State* (Chicago: University of Chicago Press, 1951).

that his view was not and could not have been based upon a full-blown evolutionary conception and that therefore his argument (and, for that matter, Maritain's as well) falls into the philosophical difficulties just discussed. Both Aristotle and Maritain are centrally involved in inductive leaps in the direction of the "essence" of man while the empiricist stands by with his deadly logical hatchet slashing the bridge to pieces.

11

But the evolutionary perspective, particularly in the hands of contemporary students of animal behavior, dulls the hatchet, or more properly perhaps, puts the empiricist in such a position that his wild swings are likely to cut off his own legs. For careful observation and fieldwork, the mainstays of applied empiricism, show beyond much doubt that in an important sense of the term there are "by nature" a good many kinds of "political" animal. If we take the standard criteria for statehood—population, territory, government—as a guide, it would not be unreasonable to suggest that animals who live in groups, occupy and defend a certain territory, and exhibit a hierarchy of dominance inside the group could be called "political" animals. I am, of course, using the term "political" very loosely here. We are, after all, still trying to decide how it ought reasonably to be employed.

The fact is that these characteristics—group life, defense of territory, and dominance hierarchy—are in evidence in one form or another throughout a wide portion of the animal kingdom. It is very tempting, therefore, to make some sort of general statement such as: In the higher animals, and especially the primates, a style of life which can be called "political" is present. In so doing, however, we must mind our philosophical p's and q's. What we are saying is simply shorthand, an economical mode of expression designed to establish perspective; it is not a general law of behavior. Nothing, strictly speaking, can be deduced from it. I remarked some time ago that the ethologists—the students of animal behavior—do not typically fall into the universal generalization model of explanation and, thus, have little difficulty with what we described earlier as Easton's Paradox. The reason is simply that these scientists operate within the pattern of explanation which is appropriate to their subject matter, namely, the evolutionary pattern.

Robert Ardrey, who has done us all a great service (whatever one may think of his particular conclusions) by bringing together the findings of the ethologists, the palaeontologists, the geologists, and the evolutionary biologists in his two books *African Genesis* and *The Territorial Impera-*

*tive,** makes the point as sharply as it can be made in a passage summing up his description of the findings with respect to territorial behavior:

Natural selection has through all its long history shown a mighty open-mindedness towards any new idea that works. Random mutation may present the kudu with horns like elongated corkscrews, the impala with horns like a bent-in lyre, the waterbuck with horns like a graceful pitchfork, and the gemsbok with horns like lances. None fail to perform the necessary function, so all have been tolerated. And natural selection has been no more dogmatic in the evolution of territorial character, size, or means of defence.

We have inspected the moving territory of the lion, the circular territory of the wolf, and the double territory of the hippo. The intensely territorial domestic dog defends a property coinciding precisely with his master's fence lines. Seasonal variations affect the moose. In winter he confines himself to a restricted "moose yard." In summer he expands his territory to include from three to ten square miles. Arboreal creatures such as birds and primates determine three-dimensional territories by volume. The gibbon will defend from thirty to one hundred acres according to the heights of the trees. A squirrel will defend three large trees or five small ones. Any variation in the size or character of a territory will be tolerated by natural selection so long as the variation acts in the interests of species survival. Even neutral territory, if it is to provide survival value, will be encouraged.

Antelope observe the neutrality of the water-hole. Most resident birds establish individual territories only through the breeding season, and observe the neutrality of the feeding ground throughout the winter. It serves the interest of dogs to affect two different personalities: to be a hostile belligerent on his master's territory, and an amiable tailwagger in the neutrality of the street. But the crowded conditions of seal rookeries have produced neutral territories of the most startling order. Narrow corridors of access lead from the sea to properties boasting no riparian rights. It is to the interest of the species that such corridors exist, and their neutrality is respected by every jostling bull in the rookery.

As anything goes that works, concerning the character of a territory, so anything goes that provides its means of defence. It is the male, for example, who almost invariably is the bearer of the territorial instinct, although his mate may assist in territorial defence. But natural selection has tolerated exceptions even to this all-but-universal law.

The phalarope is a water-bird related vaguely to the sandpiper and it frequents the Arctic in summer. It is a freak. Some chance mutation once affected the phalarope's ancestral line and in consequence certain sexual characteristics suffered reversal. The male is dun-colored, the female brightly

---

* *African Genesis* by Robert Ardrey. Copyright © 1961 by Literat S.A., and *The Territorial Imperative* (New York: Atheneum, 1966). See also Konrad Lorenz, *King Solomon's Ring* (London: Methuen, 1952) and *On Aggression* (London: Methuen, 1966).

feathered. The female arrives first at the breeding grounds and conducts the territorial scramble. The male arrives later and incubates the eggs while she defends the home place. The system works and evolution shrugs.*

Evolution shrugs indeed, and this makes universal statements of more than a summary nature absurd. It also presents me with a difficult problem of exposition. Describing all of the evidence relevant to understanding the ways in which animals, and particularly the human animal, are "by nature political" would require an extensive narrative which would clearly be out of place in the present context. There is at the moment no book which undertakes this precise task, but the works of Ardrey and Lorenz cited above will serve well until a more explicitly political book comes along.

12

We shall, then, although not without a certain measure of regret, leave the details of the animal behavior studies to the exposition of Ardrey and Lorenz. The point for us is a straightforward one, and the evidence for it is overwhelming. We observed a few pages ago that evolution works not only by selecting favorable anatomical and physiological characteristics but also by selecting favorable patterns of behavior. These are not separate processes of selection; they are inextricably intertwined. In the case of man the relatively defenseless body, the upright stance, the prolonged period of infancy all go together with the big brain. This combination of characteristics would have long ago perished in the abyss of time had it not been for the factor to which they are inextricably tied, social behavior. One can make a logical argument for this statement; for example, the use of the big brain implies an opportunity to learn (the prolonged period of infancy) which implies the capacity to teach which implies society. While such an argument is sound enough as far as it goes, the relevant evidence is by no means just a matter of logic. Man did not invent society to serve his particular purpose. Nature invented it long before man came on the scene to serve her own purposes.

The factors mentioned earlier—group living, defense of territory, and intragroup dominance—are to be found in varying combinations and styles throughout the higher reaches of animal development. It is fair to say, moreover, after recording the suitable note of caution, that the closer the animal gets to man in a physiological, anatomical, and evolutionary sense, the more these factors are arranged in a human way. Cer-

* From *African Genesis* by Robert Ardrey. Copyright © 1961 by Litertat S.A. Reprinted by permission of Atheneum Publishers.

tain kinds of monkeys, baboons, and chimpanzees are, as might be expected, the most comparable cases. The evidence strongly suggests, therefore, that not only can man reasonably be called a "political" animal, he is in a real sense "by nature" a political animal in the same loose sense that monkeys, baboons, and chimpanzees are political animals.

There seems to be little doubt that these characteristically "political" modes of behavior are in animals lower than man instinctively determined. It is a little hard to imagine a baboon social contract or constitutional convention. If Ardrey and Lorenz are to be believed, and their arguments and evidence are powerful, there exists, lingering in man as a part of his evolutionary heritage, in some sense a group instinct, a territorial instinct, and an instinct to dominance. These are the foundation stones of society and in a curious way these instincts are what gives modern universal-generalization social science whatever force it has.

There are to be sure in a general way certain kinds of regularities in human behavior, both individual and social. Some are derivative of biological inheritance and some of cultural inheritance, but none is a universal law from which precise predictions can be deduced. It is a philosophical irony of the grandest sort that those truths about, say, learning behavior that can be established as correct by behavioral psychology are probably the results of an instinctive pattern whose existence behavioral psychology denies on principle. Structural-functional anthropology begins by denying the relevance of evolution on methodological grounds and then proceeds to discover the functions and structures produced by evolution. While it is surely true that universal-generalization social science sometimes comes up with the right answers, it suffers from one fatal flaw. It cannot *in principle* tell a genuine regularity from a merely apparent or transitory one. When one is methodologically restricted to scratching the surface, it is impossible to tell whether the nugget was dropped in this particular place by an itinerant prospector or whether it sits atop a rich vein.

13

We have in the last few paragraphs played fast and loose with the term "political." For in the human context we surely mean more by "political" than a tendency toward group behavior, territorial defense, and intragroup dominance. If man is properly to be understood as the product of an evolutionary process, first biological, then cultural, we have touched so far only on the biological part of politics. We are now ready for the creative, cultural part. The division here between biological and cultural is only an expositional division; it is not a natural dichotomy. In this sense man is "by nature" a political animal not only in an instinctive sense, but also in the creative sense—for a culture-creating animal is the sort of thing that nature has produced under the title man.

On this subject let us quote cultural anthropologist Elman R. Service:

> Man is a vertebrate, mammalian, and primate animal. That is what man *is,* and the fact should never be lost from account. Yet there is something peculiar about man. The usual way of stating it is to say that man has culture, or that he has symbolic communication which results in culture. These are ways of saying that man himself has somehow added to the purely biological and situational determinants of his behavior certain others of his own invention which have increasingly involved his individual self-interests in a simultaneous commitment to his fellows. This is not always nor exclusively the case, of course, for man is still an animal, but so often and so strikingly is it so that much of philosophy and religion as well as several sciences have been concerned with the unique aspects of man's behavior rather than with the biological continuities which ally him with the other primates.
>
> The difficult subject at hand now is to discuss, or speculate about, the continuities and discontinuities of the man-primate relationship in the most pertinent context, the origin of human social culture. It is not "mere" speculation to do this, however, for there is information about primate social life, about early man in the archaeological record, and about early types of society retained by some ethnographically known hunting-gathering bands. These items, however sparse, can be used to temper the speculation.
>
> The phrase "human culture" is redundant. Culture is human and only human. It depends on an as yet inadequately defined mental capacity of human beings to communicate with each other and, correlatively, to think imaginatively in ways that apparently no other animal can. Other animals communicate and "think" but in no case can it be shown that they relate future times, other places, and even nonexistent things and places with each other. This mental gymnastic has been called the "symbolic capacity" by the ethnologist L. A. White (1949) and has been discussed in other ways by linguists (as grammar, for example, by Greenberg, 1959) and by philosophers (especially Cassirer, 1944). For the present purpose, the salient feature of man's symbolic capacity is that with it he socially creates determinants of his own behavior; that is, he invents and communicates cultural rules and values which influence his social life. This is the point at which certain subhuman abilities and propensities are emphasized, submerged, or altered, and discontinuities between subhuman primate and human behavior arise.
>
> It sounds paradoxical, or perhaps illogical, to describe culture as a determinant of the behavior of the very species that created it. But if we distinguish between the origin of a trait and its later symbolic existence as a part of an ongoing cultural tradition, then we can speak of it in this latter phase as constituting a determining factor (among others) in the behavior of every new member born into the society. It is a part of the "social heritage." Thus the statements "Man creates culture" and "Culture creates man" may be equally apposite, though opposite, generalizations.*

* Elman R. Service, *Primitive Social Organization* (New York: Random House, 1962), pp. 34–35.

We have earlier described cultural evolution as a socio-genetic system, that is, as a social system for the transmitting of information from generation to generation. Society as a survival mechanism necessarily involves the transmission of a good deal of complicated information, for each individual must somehow be carefully instructed as to his proper role if the advantage of social life is to be exploited. Nature seems to have begun the use of society at the genetic level, elaborately instructing individuals by means of instinct.

The insect group which branched off from the tree of life at a much earlier stage than the mammals has over time developed extremely complicated forms of social behavior purely on the basis of instinct. For reasons discussed earlier, the big brain as a tool was not available in the passageway of the insects. In the warm-blooded line, however, the physiological device called the brain has made possible a combination of instinct and learning. Investigation of all of the possible and all of the actual combinations of instinct and learning has just begun, but it is nonetheless possible to conclude that the larger and more complicated the brain the greater is the reliance on learning. Again, I speak only in shorthand. Evolution works pragmatically and not in accordance with the human inclination for neatness and generality.

Ardrey reports an experiment conducted by the South African naturalist Eugene Marais. Marais separated an infant otter and an infant baboon from their fellows and their natural environment for three years. He then returned each to its natural state. The otter, although he had never before seen a body of water, hesitated for a few seconds, plunged in, and shortly caught a fish. The baboon on the contrary was helpless, frightened in fact of his natural food, and had to be rescued from disaster by Marais.

Man is learner *par excellence* and the evidence suggests that he built culture upon the patterns established by his primate ancestors who relied more heavily, but certainly not exclusively, on instinct.* According to Service:

> The acquisition of culture depended upon a development of the primate brain to the point which made possible the use of symbols in communication and thought. With symbols humans can plan ways to cooperate and create means to enhance and perpetuate the cooperative relationship. . . . Once symbolic thought and communication become possible new determinants of behavior can be invented on the basis of evidence or knowledge which is already present. Sanctions, rules, proscriptions, and values can be created and established which inhibit conflict and strengthen solidarity. . . . The few

* Ardrey's report of the evidence concerning *australopithecus africanus,* the extinct man-ape of the African savannahs is most illuminating. See Ardrey, *African Genesis.*

forms of social dependence found in ape society could, by cultural means, be greatly extended and intensified. Data from all known human groups attest to the enormous importance of sharing as a means of creating friends and allies or of strengthening existing amiable relationships. The more primitive the society and the more straitened the circumstances, the greater the emphasis on sharing, and the more scarce or needed the items the greater the sociability engendered. All that was necessary, then, was the symbolic ability to make some rules and values which would extend, intensify, and regularize tendencies which already existed.*

14

Because nature is pragmatic, it does not follow in any simple way that man is also pragmatic. Freud's suggestion that nature's production of the human being was a "careless and uneven piece of work" has to be taken seriously if somewhat poetically. Freud seems to have carried *in his mind* a standard of perfection, of neatness, of regularity that in his judgment nature did not meet. But, as we have argued, nature possesses no such standard of perfection or regularity. Only the human intellect demands neatness and with this statement Freud demonstrates a powerful aspect of his humanity.

I take it, however, that this is not what Freud intended to point out when he made the statement. What he meant to suggest is that nature did not provide men with basic equipment fitted precisely to his needs. Freud being Freud we can be sure that he was particularly struck by the fact that the human sex drive was far more powerful and pervasive than it needed to be for the task of reproduction. The whole of Freudian analysis follows from this premise and who can deny that he was in some measure correct.

Konrad Lorenz makes a similar point with respect to aggression. Animals, Lorenz argues, whom nature has provided with powerful weapons have also developed inhibitory instincts which prevent destruction of the species by the species itself. Lorenz contrasts wolves and doves. Wolves, powerfully equipped for bloodletting, while they engage in frequent contests for dominance almost never fight to the death. A dominance contest between two male wolves always ends with a ritual. The loser positions himself in a certain way relative to the winner and deliberately exposes his throat, the most vulnerable spot on his body. The winner snarls and nudges, but *he does not bite*. Dominance contests among doves imprisoned by man so that the loser cannot retreat result in carnage of a most brutal and final sort. Lorenz concludes that man has a special problem. Whatever

* Service, *Primitive Social Organization*, pp. 41–42.

inhibitory instincts he has do not and cannot mitigate man's aggression when that aggression is manifested through extensions of the human body, namely, weapons whether they be bows and arrows or hydrogen bombs. Thus, again one might say that evolution's combining aggression with the big brain and the opposable thumb was an uneven and careless piece of work.*

Whether he satisfies our inclination for neatness or not, whether we understand him as an uneven and careless piece of work or not, from the point of view of evolution the human animal has "worked," that is, he has survived, which from evolution's point of view is the only thing that counts. What about the intellectual faculty itself? Do we really want to subscribe to the notion that this piece of equipment is some sort of ethereal spirit unattached to nature? What plausible reason could there be for such a bizarre conclusion?

Claude Lévi-Strauss begins his book *The Savage Mind* ** with an extensive cataloguing of the inclination of primitive peoples to engage in the most elaborate classification of the world around them. He cites the reports of a wide variety of observers in different parts of the world. Lévi-Strauss, by way of example, quotes the observations of a prominent ethnologist studying the Indians of the northeastern United States and Canada who emphasizes the wealth and accuracy of the Indians' zoological and biological knowledge and then continues:

> Such knowledge, of course, is to be expected with respect to the habits of the larger animals which furnish food and materials of industry to primitive man. We expect, for instance, that the Penobscot hunter of Maine will have a somewhat more practical knowledge of the habits and character of the moose than even the expert zoologist. But when we realize how the Indians have taken pains to observe and systematize facts of science in the realm of lower animal life, we may perhaps be pardoned a little surprise.
>
> The whole class of reptiles . . . affords no economic benefit to these Indians; they do not eat the flesh of any snakes or batrachians, nor do they make use of other parts except in a very few cases where they serve in the preparation of charms against sickness or sorcery.†

* Lorenz's suggestion that man is unique because he alone practices intraspecific aggression, that is, kills his own kind, is somewhat more dubious. It makes too much of the fact that man is biologically one species. Given the dominance of the cultural component in human behavior, we should perhaps think in terms of cultural speciation (speciation by language and cultural group) in terms of which man probably doesn't act much differently than any other animal.

** Published in the United States by University of Chicago Press. Copyright 1967 by University of Chicago Press.

† F. G. Speck, "Reptile Lore of the Northern Indians," *Journal of American Folklore,* vol. 36, no. 141, Boston-New York, 1923, p. 273.

And yet, "the northeastern Indians have developed a positive herpetology, with distinct terms for each genus of reptile and other terms applying to particular species and varieties." *

Lévi-Strauss, after listing a variety of examples comparable to the one just quoted, goes on to present an extensive list of cures and remedies developed by primitive peoples. He then concludes:

> Examples like these could be drawn from all parts of the world and one may readily conclude that animals and plants are not known as a result of their usefulness; they are deemed to be useful or interesting because they are first of all known.
>
> It may be objected that science of this kind can scarcely be of much practical effect. The answer to this is that its main purpose is not a practical one. It meets intellectual requirements rather than or instead of satisfying needs.
>
> The real question is not whether the touch of a woodpecker's beak does in fact cure toothache. It is rather whether there is a point of view from which a woodpecker's beak and a man's tooth can be seen as "going together" (the use of this congruity for therapeutic purposes being only one of its possible uses), and whether some initial order can be introduced into the universe by means of these groupings. Classifying, as opposed to not classifying, has a value of its own, whatever form the classification made takes. . . . The thought we call primitive is founded on this demand for order. This is equally true of all thought but it is through the properties common to all thought that we can most easily begin to understand forms of thought which seem very strange to us.**

15

The socio-genetic system of information transmission seems then to have not only its "genes," its individual ideas and pieces of information, but also its "chromosomes," its vehicles for the carrying of groups of ideas and pieces of information. The inclination to order, to classify—Lévi-Strauss suggests—is built-in. It is an important part of the way that the big brain operates in dealing with its environment. We have no reason to be surprised by the fact that nature would select for the capacity to learn and to teach when confronted with the peculiar characteristics of the human animal. And the capacity to learn and to teach involves the ability to organize information, to generalize it, and thus to pass it on efficiently. We have, further, no reason to be surprised when Kuhn and Toulmin call attention to the importance of paradigms and ideals of nat-

* Levi Strauss, *The Savage Mind*, p. 8.
** *The Savage Mind*, pp. 9–10.

ural order or to be surprised at Aristotle's acceptance of the premise of order. For Aristotle is a man like those that we have come to call primitive, and so even are modern scientists.

We all know, when we reflect on it, that uncertainty and inconsistency fill us with the same sort of vague anxiety induced by failure in achieving status or rejection by the opposite sex. Reflect on the universality of religion or on the easily observed fact of contemporary American life that frustration breeds the fanatical, consistency-at-all-costs political understanding of the senile Old Right and the infantile New Left. The inclination to "put things together," to achieve a consistent, universal view is part of man's natural state, and when it comes to seeing the world as it is we may perhaps note another aspect of Freud's notion of the "uneven and careless piece of work."

Ashley Montagu describes nonliterate man and his intellectual processes:

> What happens is reality to the non-literate. If ceremonies calculated to increase the birth of animals and the yield of plants are followed by such increases, then the ceremonies are not only connected with them but are part of them; for without the ceremonies the increase of animals and plants would not have occurred—so the non-literate reasons. It is not that the non-literate is characterized by an illogical mind; his mind is perfectly logical, and he uses it very well, indeed. An educated white man finding himself suddenly deposited in the Central Australian desert would be unlikely to last very long. Yet the Australian aboriginal manages very well. The aboriginals of all lands have made adjustments to their environments which indicate beyond any doubt that their intelligence is of high order. The trouble with the non-literate is not that he isn't logical, but that he applies logic too often, many times on the basis of insufficient premises. He generally assumes that events which are associated together are causally connected. But this is a fallacy which the majority of civilized people commit most of the time, and it has been known to happen among trained scientists! Non-literates tend to adhere too rigidly to the rule of association as causation, but most of the time it works. . . .*

16

We ought now to be able to see that the scientific culture which we in the contemporary West take so much for granted, which in fact we take to be the intellectual standard of the universe is an abnormal and unusual state of affairs, a special case, when viewed against the 16,000 years or so of distinctly human development. The method of science is a set of rules

*Ashley Montagu, *Man: His First Two Million Years* (New York: Columbia University Press, 1969), p. 200.

built up over time as a way of checking our natural tendency to over-generalize. The substance of scientific achievement is intellectual creation (but so are art and metaphysics, and story-telling intellectual creation), but it is intellectual creation limited and checked by rules designed to discover error. Accounts of science which stress verification and those which stress creative imagination are both partly right. They become wrong only when they attempt general characterizations which purport to define science and when they fail to notice that science occurs in time.

We are faced with the fact—not the speculation—that scientific culture, industrial society, and modern politics have developed in the West since 1500 A.D. These phenomena are to be found nowhere else (save by trans-plantation) and never before. Why? This is the question which no matter how complex and how difficult must be faced even if we do not know enough to give a complete answer. Many signs of caution and an equal number of uncertainties warn us to set ourselves a meaner question; but no less a question will do, and we are obliged, therefore, to be bold.

Recall now my earlier question about Aristotle. What was it about his cultural circumstances that allowed him to see man simply as by nature a political animal while we moderns on the contrary see so many philo-sophical holes in this statement? This, I think, is a variation on the broader question, for Aristotle was not a modern man; he did not live in the sci-entific culture. The question is for us a particularly pertinent variation for our interest is politics.

It has already been suggested in a variety of ways and from a variety of angles that cultural evolution is a matter of information transmission, a socio-genetic system we have called it (following Waddington). Let us take our bold leap at the questions before us from this piece of solid ground. If, then, cultural evolution is a matter of information transmission, then what information is transmitted and in what form it is transmitted ought to be crucially important for cultural change. I mention "what in-formation" is transmitted and "in what form" the information is trans-mitted as if there were two distinct things involved. The fact, however, is that the content and the form in which it is presented and received are inseparable parts of any transmission of information. If I wish to transmit to you a notion of the bird which I now see at my window, it will make a great deal of difference if I present the printed words "the red cardinal," a photograph, a painting, a sketch, or if I am able to say the words to you or even sing them. Indeed, there are many passages in this book which I wish that I could *say* to you so that you could hear the tentativeness or the emphasis that my voice would carry.

Let me compare two quite different kinds of "information receiving" experiences. Imagine yourself, first of all, attending a church service, preferably one of the more ceremonial sort, say, Roman Catholic, Episcopalian, or Lutheran. I have something in mind here like the state funeral of Sir Winston Churchill at the moment of the singing of "The Battle Hymn of the Republic." Put yourself, for the second situation, in a classroom taking an introductory course in symbolic logic. If you have missed that experience, an algebra class will probably do just as well.

In the first case you are virtually surrounded by the communication. The complex of tones that makes up the music rings in your ears. The pew is hard and smooth and it vibrates with the thunder of the organ. The light plays on the stained glass, the golden candlesticks, and is absorbed by the black vestments. You sense the congregation, its breathing, its occasional cough. All of your senses are involved and you are in touch with your ancestors and their practices in a hundred ways at once.

But the situation in the logic class is precisely focused. All that counts are the marks on the blackboard and the way in which they represent a theoretical reality. There is no smell, no feel, no taste, and the sound of the instructor's voice is profoundly subordinate to the marks on the slate. We may like to say that pure reason is working here, but notice that reason is working through the visual sense, and to all intents and purposes the visual sense alone, isolated from all the other senses. Is it, perhaps, that this is what makes us incline to regard the latter case as one of *pure* reason and the former as mixed, even inundated, with emotion? What after all do we mean by "pure" in this context? If it occurs to you, as it did to me when I began to write these examples down, that it is rather peculiar to call the church service an instance of information transmission, reflect on the possibility that our modern minds have given the term "information" a bias toward situations of the logic class type. Did "knowledge" or "wisdom" mean the same thing to Plato or Aristotle that "knowledge" means to a twentieth-century Logical Positivist? If not, why not?

We have talked about the social transmission of information as the core of cultural evolution. There must, we have said, be a teacher-learner relationship. Prolonged infancy and the big brain make possible a sociogenetic system for information transmission. All of this is hard to deny and yet it does not touch the question of how culture evolves. For it is significant not only that culture moves through time, is passed on from generation to generation, but also that its content has changed radically, especially in the last few thousand years.

The question, of course, is how are we to explain the changes? A great many answers to this question or to ones quite like it have been offered. For Huntington it was climate, for Hegel the march of the Absolute Spirit, and most plausibly perhaps, for Marx the relationship of man to the means of production. We are all, I am sure, familiar with the kinds of criticism advanced against these various proposals. Because, however, answers have typically been overstated—in terms of *the* key to history—this does not detract from the overriding importance of the question. Notice that the typical explanation tends to lay stress on changes in the environment to which men react in a new way. But culture is not something which occurs in the physical environment—a climatic disturbance is not culture—it is something which occurs in the process of communication between men and between generations. There can be little doubt that the invention of the steam engine made a great deal of difference to the modern West, but saying this does not speak to the question of what circumstances would make a man want to invent a steam engine, know what he had done, and want to communicate this information to others. And even more importantly, what circumstances allow or persuade a man to want to invent anything at all? So far as we know, Buddhist monks have never been interested in inventing anything whatsoever.

Lately we have begun to hear a line of argument which stresses the impact of changes in the form and style of the cultural communication process itself.* An alteration in the physical environment cannot become culturally significant unless it is in fact reacted to, grasped, and communicated by man—and any understanding and communication of that understanding can only be done through the means of communication available. If we look back over the development of Western man we can discern stages in the development of means of communication. They overlap and merge in various ways, but stages are recognizable nonetheless.

Earliest communication was undoubtedly almost entirely oral, supplemented, of course, by gestures. This oral period is by far the longest, constituting by definition the whole of human prehistory. Then pictographic or ideographic writing developed, first on stone and clay and later on more portable substances. This was presumably an extension of early graphic representations like those which have been discovered in the caves of Spain. For the West the next important step was the development of a phonetic alphabet by the ancient Semites. This, it is interesting to note and it may indeed be profoundly significant, is a step which the Chinese never took. The Greeks grasped the phonetic alphabet, adapted it to their fertile oral tradition, and the West has never been the same since. There follows an extended period in which culture was carried by the copying, recopying, and reading aloud of manuscripts. The portability of written messages and regulations literally made the Roman Empire possible. The fact that Christianity not only survived but conquered the barbarians is closely related to Christianity's control of information through its nearly

---

* See especially Harold A. Innis, *Empire and Communication* (Oxford at the Clarendon Press, 1950), *The Bias of Communication* (Toronto: University of Toronto Press, 1951), and Marshall McLuhan, *The Gutenberg Galaxy* (Toronto: University of Toronto Press, 1962).

exclusive possession of the ability to deal with manuscript. And then, at the dawn of the modern era came Gutenberg and the printing press. I shall, for the moment at least, leave the development of radio and television to Marshall McLuhan who deals with it so imaginatively.

That the printing press created the public and the possibility of mass politics is, I think, so obvious that in one sense it scarcely needs discussion. And it certainly has not been discussed at least by social scientists. Western society and its politics—and preeminently that of the United States—grew up with the printed word; but because we have not taken time and cultural evolution seriously, we have built up elaborate accounts of Western politics and then been shocked to discover that they did not apply to the essentially oral cultures of Africa or the manuscript cultures of Asia.

I shall say more about these matters in the next chapter. Let us now return to the problem of classical and modern, of Aristotle and Hobbes. When we talk in broad categories such as classical and modern, we are, as our earlier discussion has clearly indicated, dealing with all-encompassing matters of perspective, with fundamental modes of thought. My contention, which follows upon the insights of Innis and McLuhan, is simply that the dominant means of information transmission conditions, primarily by its form, the dominant mode of thought. A good many writers on epistemology have accepted the notion of Kant that the human world is defined and delimited by the structure of human thought. What they have not typically noticed, however, is that the structure of human thought has not always been the same, that the very structure of thought is affected by the way in which information moves from person to person.

The culture of the Athens of Socrates, Plato, and Aristotle was essentially oral, or more broadly what McLuhan calls audile-tactile. From the point of view of political theory the center of attention was the polis and the polis was understood as a natural and an oral phenomenon. It was speech (not writing), according to Plato and Aristotle, that opened the possibility of fullest human development. Socrates wrote nothing, Plato presented his ideas as speech transferred to writing (the dialogue form), and Aristotle's "writings" are said to be notes taken by his students.

Aristotle did not receive information by poring over the printed pages of an Encyclopaedia Britannica. He was not, and could not have been, the detached observer *looking* out at the world through the medium of the printed, and thus wholly visual, proposition. The world was all around him—he heard it, he felt it, he saw it, he smelled it, and all of these at once. Ethnologists of almost every description and opinion stress the unity of nature in the primitive mind, the lack of distinction between subject and object. We quoted Lévi-Strauss a few moments ago noting that the important thing from a primitive point of view was whether or not there was a perspective from which a woodpecker's beak and a human tooth could be seen as "going together." We must recognize that Plato and Aristotle for all their towering achievement were from an evolutionary point of view just a step or two from the primitive. Should we then be surprised that Plato assumed that because there was a noun "justice" that there must be something in reality that corresponded to it.

You may very well be thinking something like this: "Aristotle was not the only one who had the world all around him. I smell, feel, hear, and touch things as well!" But think a little further. Our whole culture is built upon seeing behind the appearances. We are *taught* and we *learn* to realize that the roar of the automobile is incidental to the chemistry of gasoline combustion and to the physics of the piston rod. Aristotle did not think by holding a printed proposition before his mind and seeing how it relates to reality, but we do. Learning for us simply *is* a matter of mentally taking something apart and seeing how it "works." This does not encompass all that we learn and know, but it is its core. Print and all that goes with it is the dominant form of information transmission for us and it creates the dominant mode of our thinking. This was not so for Aristotle. Suppose that all of your experiences were like church services and none of them like classes in symbolic logic. Would you think differently or not?

Descartes and Hobbes were men of genius and they caught early the implications of the new way of thinking implicit in the printed, visually dominant, method of information transmission. Descartes and Hobbes responded quickly and began to think *objectively* in a way that only makes sense in the modern world. How can I prove that I exist? What kind of weird question would this be to a man swallowed up in nature? What would man be like in a completely pregovernmental situation, in a state of nature? What sort of strange question would this be for a man to whom man was by nature a political animal?

In this sort of context summation is dangerous, but let me nonetheless attempt a crystallization that will allow us to proceed. Modern society and modern culture—that which began in Europe around 1500 A.D.— is built around the detached observer. It is only the detached observer who can attempt to conceive natural processes as a whole, who can consider a manufacturing-marketing process as a whole with the interworking of all of its parts, and who can conceive the governing of a large number of people over a large territory as a whole problem. Conceiving a political system as a whole and attempting to build one in the style of Hobbes or Madison requires a position of detachment like a man observing an ant colony.

Print, which concentrates information transmission into the visual sense, creates this position of detachment. Consider the forest fire sequence in

Walt Disney's *Bambi*. What would it be like if this had to be transmitted to us through the sense of touch or smell? I defy you to be detached from a forest fire communicated by the sense of touch, but because *Bambi* comes to us visually, we can be detached, we can consider the action from the outside.

Aristotle operating as a man in nature is not bothered by the jump from "some" to "all" involved in the proposition, "Man is by nature a political animal." He hears it, he feels it, he sees it, all at once and thus he *knows* it. We, however, in our detached position hold the *words* before our "mind's eye" and we immediately see the difficulty in moving from "some" to "all." Aristotle was not a fool—he knew the difference between "some" and "all"—but because he understood the world from another perspective the difference did not dominate his whole mode of thinking.

17

The phenomenon of politics, then, is not something for which we can provide an any-time-any-place definition. It occurs in time. It is rooted in our biological nature and it has evolved as culture has evolved. It involves instinct and natural human inclination, but it also involves creative thought and thought has not always been the same. Neither is it the same throughout the world. In 1969 we can be conscious of print culture politics rubbing against oral culture and manuscript culture politics. We can perhaps begin to make sense of our politics—that is, the politics of we humans and not simply we Westerners—if we tear ourselves loose from print culture science and begin to look at man in the whole sweep of nature.

We cannot begin to say all that can be said let alone all that needs to be said. But we can say something. To this task the next chapter will be devoted.

The final chapter will take up an even more mysterious consideration. If we have moved through oral culture, manuscript culture, and print culture in the Western world, from what perspective are we now looking at the world and at politics?

9
Toward "Cultural DNA": Political Evolution and Information Transmission

1

Constructing a comprehensive account of the evolution of politics is, of course, a task far too large for me to attempt here. What I propose to do in this chapter, then, is simply to outline the task and to discuss some of the issues which are likely to arise in such an enterprise.

We must from the outset be clear about the fact that political orders do not evolve independently. They are inextricably tied to broader cultural developments and seriously treating them as independent for other than expositional purposes would be a profound error. As was suggested earlier, insofar as man is by nature (or more precisely perhaps, by evolution) a social animal, he is also a political animal. Thus, any account of the development of political orders must have a core which is biological and anthropological. When, however, we speak of the evolution of politics as an aspect of cultural evolution, we are dealing with a species of human artifice. Change in a political order occurs when men react to a change in their environment according to their perception of the environment and the changes in it. And the nature of perception, as we have already noticed in a wide variety of contexts, is substantially determined by the cultural tools available. Thus, the mechanical paradigm was available to Newton and Madison in a way that it was not to Aristotle or Confucius. A man in politics whether he be theorist or actor or both will react to problems according to the way in which his cultural equipment allows him to perceive them. I do not by these remarks mean to suggest a rigid determinism anymore than a geneticist would suggest that mutations are determined. On the other hand it is scarcely conceivable that a tadpole will come into the world equipped with a pair of feathered wings.

The relationship between the evolution of political orders and the evolution of culture is not at all simple. We can begin by suggesting a case in which a change external to culture (for example, in the physical environment such as a substantial alteration in climate or an earthquake that brings a river to the surface) produces a general culture change which in turn affects the political order. This is straightforward enough, but we cannot leave the matter here. The cultural situation prior to the coming of the environmental change will affect the perception of the change, so that two cultures might react quite differently to the same sort of external alteration. The implication of the foregoing statement is quite important, for what it shows is that technological change is not external change but cultural change. Technology is human artifice and thus it is part of culture, determined and produced by standards or modes of perception that are also part of culture. Put this way the point may seem an obvious one. It is important, however, to be explicit about it, because the notion that technology (interpreted broadly to include all of the ways in which man uses the physical environment) is part of culture flies in the fact of what might be called vulgar Marxism. Karl Marx was no doubt a subtler and deeper man than many who followed him, but even he, insofar as he suggests that culture is but a superstructure built upon a foundation of the means of production, builds upon a notion that is profoundly wrong.

It is easy enough to understand the appeal of what can be described as the materialist interpretation of cultural evolution whether it be in the form of some variant of economic determinism or the more sophisticated notion of some twentieth-century anthropologists who suggest that cultures evolve toward greater utilization of energy. Such pronouncements have the ring of science and the scent of moral neutrality about them and this no doubt adds to their persuasive leverage. Technology is beyond any doubt basic and central to cultural evolution, but it is not external. It is a mistake therefore to treat technology as if it were *an* or *the* external cause of cultural change.

Give some careful thought to the following passage from Waddington:

> For the content which in man is passed from one individual to another by the use of language symbols it is conventional and convenient to use the word "culture." This word has passed through many vicissitudes, even in the recent past. Sometimes it is employed in a restricted sense, to apply only to what we regard as the higher flights of civilization but this usage is becoming less common at least in scientific circles. Archaeologists use it with particular emphasis on the material possessions—the "material culture"—of a people. But it should be noted that the material artefacts of a human society are by no means independent of the content of their verbal interchanges. This is well brought out by considering the definition of man which is generally accepted by palaeontologists and students of stone implements. Man, they say, can be described as the primate which makes tools of definite and standardized patterns. Every part of this definition is necessary. The criterion must be tool-making, not tool-using, since many lower animals (for instance, one of the species of Darwin's Galapagos finches or the California sea-otter) may use natural objects as tools. Again, some non-primates make artefacts which can be regarded as tools and which are of standardized patterns, for instance, bird's nests. Finally, some primates, such as apes, may make tools but to non-standardized patterns, such as by fitting sticks together or breaking a stick to convenient length. But man makes tools to a standardized pattern; and the standardization is not laid down through inherited instinct, as in birds, but is transmitted to him through the mechanisms of social communication. *Thus, his material culture is a manifestation of the content of the communications being carried on in his society.**

We can say then, as indeed I strongly suggested in the preceding chapter, that the mode and content of communication is at the center of culture and therefore of any account of cultural change. If this is true, as Waddington argues, of material artifacts, then how much more true must it be of politics which, after all, is a predominantly verbal activity. Before pursuing the implications of this observation in greater detail, it is necessary to make an important distinction with respect to the nature of the evolutionary process itself.

2

In a book called *Evolution and Culture*** edited by anthropologists Marshall D. Sahlins and Elman R. Service attention is very acutely called to two aspects of the evolutionary process: the Specific and the General. The distinction can, by way of introduction, be clearly made in the area

---

* C. H. Waddington, *The Ethical Animal* (New York: Atheneum, 1961), pp. 146–147. (Italics added.)
** Ann Arbor: University of Michigan Press, 1960.

of purely biological evolution. The evolution of any particular species can be understood in terms of a step-by-step, mutation by mutation history of the development of that species. Thus, theoretically at least, any species, say the African elephant, can be traced back to the beginnings of life. Thinking about this sort of historical chain is thinking in terms of Specific Evolution. From a different point of view, however, it is possible to think in terms of the development of progressively more highly evolved forms of life. Mammals, for example, are more highly evolved than insects, but —and this is the important point—mammals are not descended from insects even though they presumably have somewhere in the abyss of time a common ancestor. This kind of evolutionary thinking is what the Sahlins and Service book calls General evolutionary thinking. Note that these are not separate processes, but simply two aspects of the same process. The difference is not a matter of the process itself but a matter of the way in which the process is conceived. In brief, it is the difference between thinking in phylogenetic categories and thinking about levels of development.

This distinction is at least equally, if not more, important in the realm of cultural evolution. Consider, for example, the bright light it casts on the old problem of cultural relativism. From the point of view of Specific Evolution the Australian aborigines may be as well (or even better) adapted to their particular environment as are contemporary Americans. From the perspective of General Evolution, however, this is not to say that contemporary American culture is not more highly evolved than that of the aborigines. Sahlins and Service sum it up as follows:

> . . . Specific evolution is "descent with modification," the adaptive variation of life "along its many lines"; general evolution is the progressive emergence of higher life "stage by stage." The advance or improvement we see in specific evolution is relative to the adaptive problem; it is progress in the sense of progression along a line from one point to another, from less to more adjusted to a given habitat. The progress of general evolution is, in contrast, absolute; it is passage from less to greater energy exploitation, lower to higher levels of integration, and less to greater all-around adaptability. Viewing evolution in its specific context, our perspective and taxonomy is phylogenetic, but the taxonomy of general evolution crosscuts lineages, grouping forms into stages of over-all development.*

3

The power of the universal-generalization model of thinking, which we have earlier subjected to so much discussion, is in fact so great that it has had its effect even on recent attempts to describe the stages of

---

* From Sahlins and Service, *Evolution and Culture* (Ann Arbor: University of Michigan Press, 1960), pp. 22–23.

political development. When in an ordinary context the question of political development is raised and the demand is made for a theory of political development, the quest immediately begins for a set of *universal stages* of political development. It seems to be taken for granted that the only sort of set of stages that would be acceptable would be a set that accurately described the actual chronological development of *all* societies. What this really amounts to is the bending, so to speak, of the universal-generalization model so that it includes a historical dimension. Insofar, however, as it is still the logic of the universal-generalization model that is being applied, a serious attempt to set down universal stages of development will fall prey to Easton's Paradox. Unless I am mistaken, this tendency toward the universal is to be found in two of the prominent recent attempts to talk about stages of political development, namely, Almond and Powell's *Comparative Politics: A Developmental Approach* and A. F. K. Organski's *The Stages of Political Development,** and probably elsewhere as well. Now, as was intimated earlier, talking in terms of the Politics of Primitive Unification, the Politics of Industrialization, the Politics of National Welfare, and the Politics of Abundance as successive stages in the way that Organski does, is surely a step in the right direction, but there is a fundamental logical mistake in attempting to stretch such a set of concepts to fit all specific developmental sequences. Honesty and common sense are likely to drive the analyst to papering over the cracks by suggesting, for example, that some of the recently independent nations seem to be trying to go through all four stages at once.

Please note well that in all my previous argument for taking time seriously I am attempting to say more than "Be scientific but add a historical dimension." For simply to shift universal static categories to universal historical categories is to cloud the issue, because—and this is the important point—such a shift leaves the paradigm of explanation essentially untouched. It remains universalist and does not become evolutionary. It is the evolutionary paradigm which allows us to see the difference between Specific and General Evolution, and this distinction is all important. Sahlins and Service illustrate my point with considerable clarity in the following passage:

> Distinguishing diversification from progress [Specific from General evolution], however, not only distinguishes kinds of evolutionary research and conclusions, it dissipates long-standing misconceptions. Here is a question typical of a whole range of such difficulties: is feudalism a general *stage* in the evolution of economic and political forms, the one antecedent to modern national economy? The affirmative has virtually been taken for granted in

* Almond and Powell (Boston: Little, Brown and Company, 1966) and Organski (New York: Alfred A. Knopf, 1965).

economic and political history, and not only of the Marxist variety, where the sequence slave-feudal-capitalist modes of production originated. If assumed to be true, then the unilineal implications of the evolutionary scheme are only logical. That is, if feudalism is the antecedent stage of the modern state, then it, along with "Middle Ages" and "natural economy," lies somewhere in the background of every modern civilization. So it is that in the discipline of history, the Near East, China, Japan, Africa, and a number of other places have been generously granted "Middle Ages."

But it is obvious nonsense to consider feudalism, Middle Ages, and natural economy as the *general stage* of evolution antecedent to high (modern) civilization. Many civilizations of antiquity that antedate feudalism in its classic European form, as well as some coeval and some later than it in other parts of the world, are more highly developed. Placing feudalism between these civilizations and modern nations in a hierarchy of over-all progress patently and unnecessarily invalidates the hierarchy; it obscures rather than illustrates the progressive trends in economy, society, and polity in the evolution of culture. Conversely, identifying specific antecedents of modern civilizations throughout the world as "feudalism" is also obviously fallacious and obscures the historic course of development of these civilizations, however much it may illuminate the historic course of Western culture. . . .

Feudalism is a "stage" only in a *specific* sense, a step in the development of one line of civilization. The stage of general evolution achieved prior to the modern nation is best represented by such classical civilizations as the Roman, or by such oriental states as China, Sumer, and the Inca Empire. In the general perspective, feudalism is a specific, backward form of this order of civilization, an underdeveloped form that happened to have greater evolutionary potential than others and historically gave rise to a new level of achievement. . . . It represents a lower level of general development than the civilizations of China, ancient Egypt, or Mesopotamia, although it arose later than these civilizations and happened to lead to a form still higher than any of them.*

4

There is little or no evidence to support the notion that man's biological equipment has changed very much in the last 20,000 years or so. On the contrary the searches of Freud and Jung, the physiologists and behavioral psychologists, and the recent investigations of the ethologists tend overwhelmingly to go in the other direction. Thus, so far as our own questions are concerned, it appears altogether sound to grant the existence of a biologically ordained social nature to man. It makes, it seems to me, a great deal of sense to understand Freud and Jung as having looked behind the sweetly rational picture of man presented by Enlightenment *culture* into

---

* Sahlins and Service, *Evolution and Culture*, pp. 30–33.

his biological nature. When Freud's translator, Mrs. Riviere, chose the phrase "Civilization and Its Discontents" for Freud's own "Das Unbehagen in der Kultur" (originally "Das Ungluck in der Kultur"), she presented the substance of Freud's perspective. For Freud saw mental illness as the result of the strains put on man's biological nature by suprabiological civilization, by the evolution of culture. Here again we can see another example of a tentacle of human inquisitiveness reaching out toward truth. Where the historian of science sees the successive development of paradigms of explanation, the psychoanalyst sees the stress on a relatively stable biology of the development of culture through time.

It is very tempting—indeed our nineteenth-century biases about the character of proper scientific activity pull us strongly in this direction—to present an account of politics based on the development over time of biological stresses.* Such a perspective, however, tends to make culture, including political forms, external causes to which the unconscious reacts. While the importance of the interplay of the biological and the cultural is undeniable, we cannot learn much about the political forms themselves when they are regarded as so many external factors. From an evolutionary point of view it is quite clear, as I have stressed earlier, that what we call politics is in substantial part a species of human artifice. It involves the molding of the social environment through the agency of political devices.** What we must consider, therefore, is the way in which politics as human artifice has found its various shapes and forms.

The first step in making sense of the development of political forms is to view them against the background of an account of General as opposed to any particular sequence of Specific cultural evolution. According to Sahlins and Service:

> The traditional and fundamental division of culture into two great stages, primitive and civilized, is usually recognized as a social distinction: the emergence of a special means of integration, the state, separates civilization from primitive society organized by kinship. Within the levels *societas* and *civitas*,

---

* There is, of course, an impressive literature which takes this line. Cf. Freud's own *Civilization and Its Discontents* and Herbert Marcuse's *Eros and Civilization*. Likewise, the speculative social conclusions of Ardrey and Lorenz fit here.

**The much maligned tradition of legalism and institutionalism in political science, of course, knew this all along. The dialectic of ideas is such, however, that behavioralism found it necessary to deny the architectonic and the "constitutional engineering" function in politics in the name of science and in order to justify its own existence. We seem, however, to be turning the corner. See, for example, Charles W. Anderson's discussion of the impact of the liberal constitution on nineteenth-century Latin America in *Politics and Economic Change in Latin America* (Princeton: Van Nostrand, 1967), Douglas W. Rae, *The Political Consequences of Electoral Laws* (New Haven: Yale University Press, 1967) and Robert A. Dahl, *Pluralist Democracy in the United States* (Chicago: Rand McNally, 1967).

moreover, further stages can be discriminated on criteria of social segmentation and integration. On the primitive level, the unsegmented (except for families) and chiefless *bands* are least advanced—and characteristically, pre-agricultural. More highly developed are agricultural and pastoral tribes segmented into clans, lineages, and the like, although lacking strong chiefs. Higher than such egalitarian *tribes,* and based on greater productivity, are *chiefdoms* with internal differentiation and developed chieftainship. Similarly, within the level of civilization we can distinguish the *archaic* form—characteristically ethnically diverse and lacking firm integration of the rural, peasant sector—from the more highly developed, more territorially and culturally integrated *nation state,* with its industrial technology.

General progress [they continue] can also be viewed as improvement in "all-around adaptability." Higher cultural forms tend to dominate and replace lower, and the range of dominance is proportionate to the degree of progress. So modern national culture tends to spread around the globe, before our eyes replacing, transforming, and extinguishing representatives of millennia-old stages of evolution, while archaic civilization, now also falling before this advance, even in its day was confined to certain sectors of certain continents. The dominance power of higher cultural forms is a consequence of their ability to exploit greater ranges of energy resources more effectively than lower forms. Higher forms are again relatively "free from environmental control," i.e., they adapt to greater environment variety than lower forms. . . . By way of aside, the human participants in this process typically articulate the increasing all-around adaptability of higher civilizations as increase in their *own* powers: the more energy and habitats culture masters, the more man becomes convinced of his own control of destiny and the more he seems to proclaim his anthropocentric view of the whole cultural process. In the past we humbly explained our limited success as a gift of the gods: we were *chosen* people; now we are *choosing* people.*

Notice now that we are here presented with a rather loose and general *set* of stages in General cultural evolution: the band, the tribe, the chiefdom, the archaic civilized form, and the nation-state. Notice further the correlation (I want very carefully to avoid imputing any sort of simple causal relationship) between these stages and the dominant mode of information transmission. The band, the tribe, and the chiefdom are stages differentiated by the anthropologist within the general category primitive and the socio-genetic system is oral. The so-called archaic category obviously encompasses a wide diversity of particular instances—everything from the civilization of the Tigris-Euphrates, to the Roman Empire, to nineteenth-century China—but remember the authors are talking about General evolution. What unites these diverse examples from the point of view of Sahlins and Service is their common range of energy exploitation. But observe as well that what unites them is the fact that the dominant

* Sahlins and Service, *Evolution and Culture,* pp. 36–37.

mode of information transmission is one or another form of nonmechanized writing. The nation-state is likewise correlated with the printing press and its products. Do not misunderstand me. I am not suggesting that an oral tradition is eradicated by writing or by print (although it is more true of the latter than the former—print culture stamping out an oral tradition is very often what people mean when they use the term "modernize"), but something closer to the proposition that some form of writing is the *sine qua non* of the archaic but civilized, and that print (or perhaps another mass medium) is the *sine qua non* of a true nation-state.

Why is this correlation of General stages with means of communication important for our purpose? We are interested in the development of the forms of human political artifice and my contention is that the forms of creative politics are profoundly involved with the forms of information transmission concurrently dominant. Men react to what they perceive and they perceive through (in the sense of looking through a pair of spectacles) a form of thinking which is shaped by the style and form of communication. Wittgenstein once said, "To imagine a form of language is to imagine a form of life." Perhaps, a form of communication is a form of language is a form of life.

5

Before we begin a more detailed discussion of the impact of the means of information transmission on political evolution, we have to take pains to make clear just what is and what is not being asserted. In these matters saying what is intended with reasonable economy of expression is not easy, so, as I have before, I must ask your cooperation or at least your indulgence.

I shall make statements like "The modern state is the product of print culture." Do I mean by this statement that Gutenberg was the *cause* of the modern state in the same sense that the downward motion of my arm is the *cause* of the window shutting? Do I mean that the introduction of print into a semiliterate or preliterate culture will automatically produce a modern state? I mean neither of these, although I suspect there is more sense to the latter than to the former. What must be understood is that such statements are being made in the context of an evolutionary paradigm of understanding and not in the context of a mechanical cause-and-effect one.

The difference between the two paradigms of understanding can scarcely be better illustrated than in terms of this example. The universal-generalization social scientist automatically structures his consideration of a statement like "The modern state is the product of print culture" as "Whenever print, then the modern state" and the whole correlation-prediction, cause-and-effect conceptual apparatus then comes into play.

The geneticist Theodosius Dobzhansky faces the same kind of problem in discussing the now popular notion that life must exist elsewhere in the universe because of the probabilities involved in the astronomers' estimates that the universe contains upwards of 100 million planets. Dobzhansky begins by suggesting that probability estimates make little sense in this context because we have no idea how probable or improbable it was that life should have begun at all. The generalizers—those scientists who are in effect thinking: "the conditions for life (temperature, atmosphere, and so on) produce life"—judge that life elsewhere is probable. Dobzhansky, the evolutionist, is much more cautious because in his conceptual scheme an isolated, peculiar—indeed, accidental—event of momentous import makes much more sense. If you can see that the difference here is not a matter of evidence but of paradigms of understanding, you will understand this book. If you do not, I have not done my job well enough. Dobzhansky continues, amplifying the point just made:

> The further argument, that once life has arisen it is bound to evolve approximately as it did on earth is less defendable still. At first blush, this may not seem unsound; many of the body structures of existing organisms are adaptations which help to solve the problems of survival posed by the environment, and it may seem that natural selection should bring them into existence wherever they are needed, as it did on earth. This is not necessarily so; in the first place, an adaptive problem may not be solved at all, and the species may become extinct; moreover, and this is crucial, many adaptive problems may be solved in more than one way. The solutions which we know to have been utilized by organisms on earth cannot be guaranteed to have been either the best conceivable ones, or the ones that would occur again in the thinkable (not, however, probable) situation of the evolution occurring for the second time here, on earth.*

What is being asserted concerning print and modern politics in the West is not a matter of generalization of any sort (although some sort of generalization concerning print *could* indeed be asserted and be subjected to test in the ordinary way), but simply that Gutenberg's device brought Western culture to a point which in retrospect we can see as opening a whole new era in human activity. Dobzhansky makes the same point:

> The flow of evolutionary events is, however, not always smooth and uniform; it also contains crises and turning points which, viewed in retrospect, may appear to be breaks of the continuity. The origin of life was one such crisis, radical enough to deserve the name of transcendence. The origin of man was another. This should not be taken to mean that the origin of life or of man was instantaneous or even very swift. A process which is very rapid in a

* Reprinted by permission of The World Publishing Company from *The Biology of Ultimate Concern* by Theodosius Dobzhansky. An NAL book. Copyright © 1967 by Theodosius Dobzhansky.

geological (more precisely, palaeontological) sense may appear to be lengthy and slow in terms of a human lifetime or a generation.

The appearance of life and of man were the two fateful transcendences which marked the beginnings of new evolutionary eras. They were, however, only extreme cases of radical innovations, other examples of which are also known. The origin of terrestrial vertebrates from fishlike ancestors opened up a new realm of adaptive radiations in the terrestrial environments, which was closed to water-dwelling creatures. The result was what Simpson [G. G. Simpson, *The Major Features of Evolution* (New York: Columbia University Press, 1953)] has called "quantum evolution," an abrupt change in the ways of life as well as in the body structures. Domestication of fire and the invention of agriculture were among the momentous events which opened new paths for human evolution.*

I am suggesting here that insofar as cultural evolution is at core a matter of information transmission, then the way in which information is stored and transmitted is likely to be pivotal in understanding how human life has evolved—more important perhaps than fire or agriculture or gunpowder, although this is surely a matter open to discussion. Note again the inappropriateness of generalizing in this context. One simply cannot say—as Dobzhansky rightly points out—that given the presence of land certain fish will find a way to walk on it. Nonetheless, the evidence is strong that certain fish did learn to walk on their fins, and that once this happened the character of life was radically transformed. Similarly, I do not suggest that the introduction of print will necessarily produce modern Western consequences; only that mankind entered a new era in Europe with the coming of print, and, in the same way, we are just beginning to realize what is happening to us now that electronic communication is circling the globe.

One further point ought to be noted. Given the power of modern means of communication, no presently underdeveloped area of the world can ever develop "modernity" in quite the way that the West did just because the people of the area cannot but *know,* in some measure at least, how the West developed over the last several hundred years. Looked at as a matter of evolutionary thresholds this situation makes perfect sense. Searching, however, for general laws in this context can end only in vagueness or frustration.

6

If we wish to understand the politics of any particular human group, we must deal with a matter of Specific Evolution. If in particular we are interested in the politics of the West we must take account of the specific

* *The Biology of Ultimate Concern,* p. 50.

steps which the peoples of the West have taken, recognizing clearly that the steps are specific. We must resist the temptation to make the steps into a general theory of the political development of all Western peoples and certainly not of non-Western peoples. The Swedish people have not produced their political forms along precisely the same lines as have the English, the French, or the Spanish and the search for truth makes it mandatory that this fact be taken seriously.

On the other hand, however, the fact that political forms are related to media of communication and that they are diffused through media of communication makes it possible to discuss successive patterns within a particular cultural context. If we begin a discussion of Western political forms with the Greeks, we must notice, first of all, that the polis is the highest form of oral culture politics. We have no less an authority for this point than Aristotle himself. After having discussed the household and the village, Aristotle suggests in Book I Chapter 2 of the *Politics* that, "The final association, formed of several villages, is the city or state." "It follows," he continues

> that the state belongs to a class of objects which exist in nature, and that man by nature is a political animal; it is his nature to live in a state. . . . Nature, as we say, does nothing without some purpose; and for the purpose of making man a political animal she has endowed him alone among the animals with the power of reasoned speech. Speech is something different from voice, which is possessed by other animals also and used by them to express pain or pleasure; for the natural powers of some animals do indeed enable them both to feel pleasure and pain and to communicate these to each other. Speech on the other hand serves to indicate what is useful and what is harmful, and so also what is right and what is wrong.

As Aristotle recognized, although of course he could not have seen its significance, the polis like primitive society was built upon speech as a means of communication. It is obvious enough that the polis as a face-to-face society was limited in size and development by the limitations of speech. What is not so obvious, however, is that the form of political thinking which is exemplified by Plato and Aristotle was also conditioned by speech. Only a few scholars have grasped the explanatory power of this simple fact.* Leo Strauss perhaps sees the point which I want to make when he says that, "Classical political philosophy is characterized by the fact that it was related to political life directly." There was, Strauss suggests, for Plato and Aristotle no intermediary in the form of a written

---

* See, in particular, Eric A. Havelock, *Preface to Plato* (Cambridge, Mass.: Belknap Press of Harvard University, 1963) and, of course, McLuhan, *The Gutenberg Galaxy* (University of Toronto Press, 1962) cited earlier, and Innis, *Empire and Communication* (Oxford at the Clarendon Press, 1950), and *The Bias of Communication* (University of Toronto Press, 1951).

tradition between the thinker and politics—let alone the intermediary of the detached observer perspective that characterizes modern science. And this directness, that is to say, this "oralness" profoundly shapes the character of the political understanding.

Classical political philosophy attempted to reach its goal by accepting the basic distinctions made in political life exactly in the sense and with the orientation in which they are made in political life, and by thinking them through, by understanding them as perfectly as possible. It did not start from such basic distinctions as those between "the state of nature" and "the civil state," between "facts" and "values," between "reality" and "ideologies," between "the world" and "the worlds" of different societies, or between "the I, Me, Thou, and We," distinctions which are alien, and even unknown, to political life as such and which originate only in philosophic or scientific reflection. Nor did it try to bring order into that chaos of political "facts" which exists only for those who approach political life from a point of view outside of political life, that is to say, from the point of view of a science that is not itself essentially an element of political life. Instead, it followed carefully and even scrupulously the articulation which is inherent in, and natural to, political life and its objectives.

The primary questions of classical political philosophy, and the terms in which it stated them, were not specifically philosophic or scientific; they were questions that are raised in assemblies, councils, clubs, and cabinets, and they were stated in terms intelligible and familiar, at least to all sane adults, from everyday experience and everyday usage. These questions have a natural hierarchy which supplies political life, and hence political philosophy, with its fundamental orientation. No one can help distinguishing among questions of smaller, of greater, and of paramount importance, and between questions of the moment and questions that are always present in political communities; and intelligent men apply these distinctions intelligently.*

Notice that when Strauss speaks of "political life" he clearly does not refer to a bureaucrat interpreting a written regulation or a judge searching a statute book; he means *oral* political life. And the style of political thinking is shaped by the oral mode. Strauss continues:

Similarly it can be said that the method, too, of classical political philosophy was presented by political life itself. Political life is characterized by conflicts between men asserting opposed claims. Those who raise a claim usually believe that what they claim is good for them. In many cases they believe, and in most cases they say, that what they claim is good for the community at large. In almost all cases claims are raised, sometimes sincerely and sometimes insincerely in the name of justice. The opposed claims are based, then,

---

* Leo Strauss, "On Classical Political Philosophy," from *Social Research* (February, 1945), reprinted in my *Plato: Totalitarian or Democrat?* (Englewood Cliffs, N.J.: Prentice-Hall, 1963), pp. 154–155.

on opinions of what is good or just. To justify their claims, the opposed parties advance arguments. The conflict calls for arbitration, for an intelligent decision that will give each party what it truly deserves. Some of the material required for making such a decision is offered by the opposed parties themselves, and the very insufficiency of this partial material—an insufficiency due to its partisan origin—points the way to its completion by the umpire. The umpire par excellence is the political philosopher. He tries to settle those political controversies that are both of paramount and of permanent importance.*

While it is accurate to say that Greek political thinking was essentially oral thinking about an oral society, Greek political thinking *together with its subsequent impact* must be seen in broader terms. Greek political thinking viewed in the context of cultural evolution is a matter not only of oral communication but of the fusing of three communication factors: an oral tradition, writing on a transportable medium, and a phonetic alphabet. What I shall argue here has the logical form "These factors came together in such-and-such a way with such-and-such a result" and not "Whenever factors *a, b,* and *c* conjoin, such-and-such will be the result."

Greek political thinking was oral thinking frozen in transportable phonetic writing. As Strauss observes:

> It was only after the classical philosophers had done their work that political philosophy became definitely "established" and thus acquired a certain remoteness from political life. Since that time the relationship of political philosophers to political life, and their grasp of it, has been determined by the existence of an inherited political philosophy: since then political philosophy has been related to political life through the medium of a tradition of political philosophy.**

Philosophy for Plato and Aristotle was a matter of oral disputation, and the dialectical method of philosophizing set down by Plato in his dialogues had a profound effect on the nature of the questions asked and on the answers which seemed relevant and correct. Here we are in a sense talking about paradigms, and we are in effect suggesting that the mode of discourse affects the character of the paradigm. The notion, characteristic of Platonic thought, of moving from opinion to knowledge by a process of dialectical discussion or dialectical introspection (having, so to speak, a conversation with yourself) stands in sharp contrast with, say, the twentieth-century conception that a true proposition is a picture of reality. As historians of philosophy would put it, it is the difference (and a profound difference at that) between Classical Realism and Logical Positivism.

---

* *On Classical Political Philosophy,* pp. 155–156.
** *On Classical Political Philosophy,* p. 153.

What Strauss calls the direct apprehension of the world and the oral mode of thinking about it led Plato and Aristotle to unhesitatingly proclaim the existence of an essential order in nature. If one cannot entertain the possibility that a question such as What is justice? has no answer, then asking the question will produce an answer.

The notion of an ordered nature, powerfully articulated by Plato and Aristotle, was set down in alphabetic writing on an easily transportable medium. These writings spread, together it might be said with the Gospels and the Epistles of St. Paul later on, throughout the Mediterranean world and were ultimately relatively easily translated into other alphabetical scripts. As Strauss suggests, wherever the writings penetrated, educated men could no longer think about politics except through the traditions of political philosophy. A paradigm of political understanding had been erected (as a part of a paradigm of natural understanding which we noted earlier in discussing the history of science) and, while this paradigm might be modified and adapted in the light of changing circumstances, only a radically new form of life and of discourse could allow it to be ignored or rejected.

The Platonic and Aristotelian conception of an ordered nature became through the agency of the Stoics and Epicureans, the Roman lawyers, and finally the Christian establishment the foundation for more than 1500 years of intellectual, moral, and political authority in the Western world. Writings during this extended period came to have a power and an authority which we no longer attribute to them. Those men who possessed the monopoly on reading and writing were the same men who exercised religious and political control. Reflect on the power and authority of the Christian Scriptures, the role of Latin in the Catholic Church, the extraordinary power of endurance of the Jews united by their ancient writings, and finally reflect particularly on the dominant method of science, philosophy, and political philosophy throughout the Middle Ages. The learned men of the Middle Ages reasoned from authority; intellectual activity was on the whole disputation on the ancient texts. Think of John of Salisbury, Thomas Aquinas, Marsilius of Padua, Martin Luther—great innovators in one sense, but even they were tied to the ancient texts.

Do not misunderstand me. I am not suggesting that the means of communication, the agency of socio-genetic transmission, is *the* causal factor. Biology, climate, geography, and a wide variety of plain accident are all involved in Specific cultural evolution. But—to repeat myself—insofar as politics is a matter of human artifice, how men think is important and how men think is related to what they think with and through. Does a mathematician think with and through his chalk and his blackboard or not? Have you ever seen the placemats or the tablecloth after a mathematicians' lunch?

7

One may perhaps argue, somewhat paradoxically, that Plato *wrote* what he wrote in an attempt to save the sense of natural order implicit in the *oral* tradition from the subversive influences of contact with other human groups introduced into Greece along with writing. The insistence of Plato on "nature" over "convention" stressed by Sabine and others can perhaps be understood in this way.* What Plato objected to, looking back over the years of the Peloponnesian War and before, was the weakening of order epitomized by the Sophists which resulted from man's detachment from nature. As Harold Innis writes,

> In the fourth century Plato attempted to save the remnants of Greek culture in the style of the Socratic dialogues which in the words of Aristotle stood halfway between prose and poetry. In the seventh epistle he wrote, "no intelligent man will ever be so bold as to put into language those things which his reason has contemplated, especially not into a form that is unalterable— which must be the case with what is expressed in written symbols." . . . The written tradition had brought the vitality of the oral tradition to an end. In the words of Nietzsche, "Everyone being allowed to read ruineth in the long run not only writing but also thinking.**

Political artifice, as is well known, turned profoundly with the Romans toward law and administration. The notion of the Natural Law became explicit with the Stoics, who modified Plato and Aristotle, and the Roman lawyers. Cicero's understanding of the Natural Law and the Roman jurisprudential hierarchy of *ius naturale, ius gentium,* and *ius civile* make sense in the context of bureaucratic regulation of diverse peoples over a large area. Innis, concerned as he is with the effect of media of communication principally in two directions, those of space and time, writes:

> The bureaucratic development of the Roman Empire and success in solving problems of administration over vast areas were dependent on supplies of papyrus. The bias of this medium became apparent in the monopoly of bureaucracy and its inability to find a satisfactory solution to the problems

* See George H. Sabine, *A History of Political Theory,* 3rd ed. (New York: Holt, Rinehart and Winston, 1961), especially pp. 30–31. Havelock (*Preface to Plato,* previously cited) argues that Plato *rejected* the oral tradition in the sense that he rejected the poetic Homeric tradition. It seems to me that the arguments are not necessarily incompatible.
** Harold A. Innis, *The Bias of Communication* (Toronto: University of Toronto Press, 1951), p. 44.

of the third dimension of empires, namely time. A new medium emerged
to meet the limitations of papyrus. The handicaps of the fragile papyrus roll
were offset by the durable parchment codex. With the latter the Christians
were able to make effective use of the large Hebrew scriptures and to build
up a corpus of Christian writings. . . . [Ultimately,] the problem of the
Roman Empire in relation to time was solved by the support of religion in
the Christian church. The cumulative bias of papyrus in relation to bureaucratic administration was offset by an appeal to parchment as a medium for
a powerful religious organization. Recognition of Christianity was followed
by the drastic suppression of competing pagan cults. The attempt of em-
perors to build up Constantinople as the centre of the civilized world especially after the fall of the Western Empire in 476 A.D. by establishing a
large library and producing a code of civil law [Justinian's] created friction
with Rome and with Alexandria . . . geographical separation reinforced differences in religion and exposed the Eastern Empire to the attacks of the
Persians and in turn the Arabs.*

We should, then, not be surprised to see first the later Stoics and then
the Roman lawyers modifying the tradition of political understanding inherited from Plato and Aristotle to fit the "transmission of information
through writing" context of their times. If, in the context, one inherited a
notion of natural order, it would make sense to interpret this order as
a supreme form of law which lay behind and above the written and customary law. Nor should we be surprised to see Christianity—that offshoot
of "the people of the book"—grasp for a permanent medium which could
preserve the record of the ultimate historical event, and in so doing establish authority based on past events which would set the tone of Western
history for a thousand years. The Roman Church in an important sense
rests its authority on the preservation of the words: "Thou art Peter, and
upon this rock I build my church."

The authority, however, is not merely political, moral, and social. It is
intellectual as well, and it is this aspect which is important for us. The
authority of the ancient pronouncements preserved in written form in the
ancient languages sets up a kind of paradigm of explanation and understanding. A proper explanation, as almost a random look at the writings
of Thomas Aquinas will demonstrate, is one in which whatever reasoning
is involved corresponds and is subordinate to a pronouncement in an
ancient text. The paradigm of explanation to which I refer lies behind the
medieval notions that philosophy, including natural philosophy, is the
handmaiden of theology and that when reason and revelation seem to
disagree reason must be wrong. A proper explanation in politics, morals,
physics, or chemistry involves citing the relevant passages from the ancient

---

* *The Bias of Communication,* p. 46.

texts (whether biblical or Greek) and elucidating the meaning by reason. We may perhaps at this point ask a rhetorical question: If this sort of authority attaches to the written word when it is the exclusive preserve of ancient writers in ancient languages, what will happen if a means suddenly is found to disseminate the written word widely and freely?

In the preceding paragraph we anticipate one of the most important turns in cultural evolution, that time somewhere in the vicinity of 1500 A.D. when the world through the agency of the Europeans began the achievement of what we have come to call modern society. Bearing always in mind that we seek to make sense of forms of human thought, of human artifice, let us again follow Innis as he describes the way in which the pattern took shape. He is speaking of the situation after the fall of Rome in the west:

> The spread of Mohammedanism cut off exports of papyrus to the east and to the west. The substitution of parchment in the West coincided roughly with the rise of the Carolingian dynasty and the decline of the Merovingians. Papyrus was produced in a restricted area and met the demands of a centralized administration whereas parchment as the product of an agricultural economy was suited to a decentralized system. The durability of parchment and the convenience of the codex for reference made it particularly suitable for the large books typical of scriptures and legal works. In turn the difficulties of copying a large book limited the numbers produced. Small libraries with a small number of large books could be established over large areas. Since the material of a civilization dominated by the papyrus roll had to be recopied into the parchment codex, a thorough system of censorship was involved. Pagan writing was neglected and Christian writing emphasized. "Never in the world's history has so vast a literature been so radically given over to destruction." "Whatever knowledge man has acquired outside Holy Writ, if it be harmful it is there condemned; if it be wholesome it is there contained" (St. Augustine). The ban on secular learning gave a preponderance to theological studies and made Rome dominant. The monopoly of knowledge centering around parchment emphasized religion at the expense of law.
>
> Parchment as a medium was suited to the spread of monasticism from Egypt throughout western Europe. St. Benedict founded a monastery at Monte Cassino about 520 A.D. and emphasized rules which made the preservation of books a sacred duty. His work followed by that of Cassiodorus gave "a scholarly bent to western monasticism." In spite of these efforts learning declined in Europe. Revival came on the fringes of the West in the independent and self-governing monasteries of Ireland. Missionary zeal led to the establishment of monasteries in Scotland and northern England and early in the seventh century on the Continent. The revival gained impetus with the support of Charlemagne and the migration of Alcuin from York. England and northern France were exposed to Danish raids but European monasteries had acquired transcriptions from English codices and

supplemented them with those from Rome. Durable parchment books could be moved over long distances and transferred from regions of danger to regions of safety.*

8

Parchment became the medium through which religion established intellectual authority. As we have suggested, a style was established, a paradigm erected, which only a substantial change in circumstances could demolish. That change, of course, came in the form of mechanized printing and we must now say a word or two about the introduction of a crucial factor which helped to call the printing press into being, namely paper.

While paper was invented by the Chinese, the printing press (in the modern sense) was not. It is instructive to attend to this somewhat curious fact. "Discovery of the technique of making paper from textiles," Innis suggests, "provided a medium with which the Chinese, by adaptation of the brush for painting to writing, were able to work out an elaborate system of pictographs. A system of four to five thousand characters was used for ordinary needs 'enabling those who speak mutually unintelligible idioms to converse together, using the pencil instead of the tongue.' Its effectiveness for this purpose meant the abandonment of an attempt to develop an alphabet system." **

In the putting together of this book I have, of course, reflected a good deal on the thesis concerning media of communication which I have been attempting to present in the last several pages. No facts which I have come across persuade me of its essential soundness more than those involved in the comparison of Western civilization with that of the Chinese.

We must begin by noting that contemporary China is no mere nation-state like those of Europe; it is more properly described as a "civilization-state." † It is as if all of Europe including European Russia had been united under a common government more or less continuously for three thousand years. The Chinese with a population of comparable size with comparable linguistic differences spread over a land mass of comparable vastness with climatic differences of comparable proportions did not split up into nation-states. Neither did Chinese civilization for all its sophistication lead the world into the modern era. The Chinese invented paper and began to use it for writing in about 105 A.D. They thus were in possession of what Innis would call a space-biased medium and they used it

* *The Bias of Communication*, pp. 48–49.
** *The Bias of Communication*, p. 50.
† See the very interesting book by François Geoffroy-Dechaume, *China Looks at the World* (New York: Pantheon Books, 1967).

in a way comparable to the Roman use of papyrus. In addition the Chinese pictographic script was itself integrative; it pictured ideas rather than sounds, it was interlinguistic rather than phonetic.

Geoffroy-Dechaume who knows China but so far as I can tell knows nothing of Innis or McLuhan speaks to the point in the following way:

> WRITING. This fixes language and thought. Writing thus makes possible the transmission and enrichment of the spiritual heritage through the course of ages. Without writing there can be little history; without it no long-distance organization, no enduring laws are possible, there can be no continuous progress. Through language, to which it gives material form, writing remains closely linked to the biological factors of which it is, as it were, the projection. The thought and culture which it sanctions spread by its means and gradually over the whole area which is being civilized. Indeed, the emergence of writing may be said to represent, in the evolution of humanity, a threshold of consciousness comparable to the genetic mutation which gave rise to *homo sapiens*. We witness the appearance of this invention in many parts of the globe without being able to ascribe a definite origin or origins to it, any more than to the various races of mankind. Phoenician writing, which absorbed Egyptian elements, spread throughout the world supplanting cuneiform and hieroglyphic scripts and producing the Greek, Roman, Arabic, and Sanskrit alphabets, but it failed to penetrate China!
>
> . . . China invented her own form of writing, her own cities, her own manifold techniques, and her originality is thus threefold and appropriates all foreign elements, subjecting even religions—Buddhism for instance, and Marxism in our own day—to radical transformations before making use of them in the cause of her own expansion.
>
> China's writing, for thousands of years the backbone of her civilization and the ferment which gave it life, must be considered in contrast with the Roman alphabet. The latter strives today to spread gradually over the whole world, yet, as a universal vehicle, it remains split up into a considerable number of languages. The Chinese script on the contrary is interlinguistic. Close on a thousand million people will soon be able to communicate by its means, for Japan still retains access to it.*

The tremendous absorbing power of Chinese culture, which for Westerners has become proverbial, is focused in its manner of writing. Writing which pictured ideas rather than sounds united people who could not talk to one another. Moreover, it set the tone for a culture capable of absorbing foreign ideas and styles of life. Nowhere save in China have Jews been absorbed almost without a trace. The very success, at one level, of the pictographic script prevented the development of a phonetic alphabet and the absence of such an alphabet placed an enormous obstacle in the way

---

* *China Looks at the World,* pp. 67–72.

of the development of mechanized printing. Devising a machine which can deal with 26 or 30 or even 100 characters is a very different matter from devising and investing in a machine that can handle 5,000.

Paper came to Europe about 1200 A.D. through Arab intermediaries. It brought with it a reexposure to certain of the classical writings including those of Aristotle. And this, of course, made possible the synthesis of Thomas Aquinas. Paper arrived in Europe in a situation of exclusive control of knowledge by religious authority through, as we have previously noted, the medium of parchment. The new medium was quickly grasped as a tool of competition by, if you will, the upwardly mobile, the small tradesman, merchant, and craftsman. Here is part of Innis's account of the matter—note especially the final sentence:

> The monopoly built up by guilds of copyists [who followed upon monastic monopolies] and others with the making of manuscripts had its effects in high prices which in turn invited attempts to produce at lower costs. It was significant that these attempts were made in territory marginal to France in which copyists' guilds held a strong monopoly, and that they were concerned with the production of an imitation of manuscripts such as Bibles, which commanded, partly as a result of size, very high prices. In 1470 it was estimated in Paris that a printed Bible cost about one-fifth that of a manuscript Bible. The size of the scriptures had an important effect in hastening the introduction of parchment codex and in turn the introduction of printing. The feudal divisions of Germany provided an escape from the more rigid central control of France.
>
> Profits were dependent on exact reproductions of expensive manuscripts. It was necessary to develop arrangements by which type could be cast resembling exactly the letters of the manuscript and in sufficient quantity to facilitate setting up pages for printing. The alphabet which had been conventionalized to a limited number of letters used in innumerable combinations in words lent itself to adaptation to mechanical production of large numbers of the same letters which could be put together in the required combinations. In contrast with China, where the character of the script involved large-scale undertakings supported by governments, the alphabet permitted small-scale undertakings manageable by private enterprise.*

We are now in a place to pause for a moment to reflect on the complexity of the mechanism of cultural evolution. The introduction of paper, the use of a phonetic alphabet, the presence of a tradition-oriented system of authority, and an adequate metals technology were all necessary to the development and success of the printing press. As a matter of Specific Evolution there is beyond any doubt a considerable measure of randomness

---

* From Harold A. Innis, *Empire and Communication* (Oxford: Clarendon Press, 1950). By permission of Clarendon Press, Oxford, p. 173.

in this whole process. Once the printing press has arrived, however, a new phase in General cultural evolution has opened, and Europe begins to lead the world into what we now call modernity. Are modern science, technology, and education conceivable without print? Surely not. Is a modern economy possible without print? Can one conceive of the growth of European nation-states around a common language and literature under the promulgation of law over a considerable space without print? Could mass democracy as an idea be taken seriously without print? Would the industrial revolution have been possible without print? Could man possibly have developed a functionally specific society without print? To all these questions and many more like them I think we must answer, "Of course not!"

When Sir Henry Maine describes modernity as the transition from status to contract, one may fairly ask, "What lies at the core of this change?" Daniel Lerner in *The Passing of Traditional Society* * describes the premises of his account of modernization in the Middle East:

> It is a major hypothesis of this study that high empathic capacity is the predominant personal style only in modern society, which is distinctly industrial, urban, literate and *participant*. Traditional society is nonparticipant—it deploys people by kinship into communities isolated from each other and from a center; without an urban-rural division of labor, it develops few needs requiring economic interdependence; lacking the bonds of interdependence, people's horizons are limited by locale and their decisions involve only other *known* people in *known* situations. Hence, there is no need for a transpersonal common doctrine formulated in terms of shared secondary symbols—a national "ideology" which enables persons unknown to each other to engage in political controversy or achieve "consensus" by comparing their opinions. Modern society is participant in that it functions by "consensus"—individuals making personal decisions on public issues must concur often enough with other individuals they do not know to make possible a stable governance. Among the marks of this historic achievement in social organization which we call Participant Society, are that most people go through school, read newspapers, receive cash payments in jobs they are legally free to change, buy goods for cash in an open market, vote in elections which actually decide among competing candidates, and express opinions on many matters which are not their personal business.**

Lerner suggests that these characteristics of modernity follow from a capacity for empathy—the possibility of an individual imagining himself in another's place. The question which our evolutionary paradigm leads us to ask is, "How did this capacity for empathy come about?" Lerner

* New York: The Free Press of Glencoe, 1958.
** *The Passing of Traditional Society*, pp. 50–51. (Italics in original.)

himself speaks to the point, "Radio, film and television climax the evolution set into motion by Gutenberg. The mass media opened to the large
masses of mankind the infinite *vicarious* universe." * Radio and motion
pictures along with print carry modernization around the world as any
contemporary student of the developing areas will testify, but print began
the process for all mankind and this for the moment is our point.

9

Our evolutionary paradigm with its stress on information transfer leads
us to investigate the significance of the invention of printing and to inspect
the means of its effect. We can answer with Lerner and in some ways with
Innis at what might be called the sociological level. We can, as we have
for the past several pages, discuss the possibilities which print opens in
matters of organization over space—political, social, economic—and we
can discuss the effects of print on the experience of individuals. In so
doing we are taking the detached observer position of the modern scientist.
We are, so to speak, looking down on society like men watching an anthill
or, if you will, like men watching rats in a maze. From this view—which
I called sociological a moment ago—we can see effects brought about by
the introduction of a new kind of communication.

But more is involved. Our evolutionary perspective also demands, as I
observed many pages ago, that we see the whole phenomenon of man,
that in short we inquire into the question of how we, as modern men, are
able to adopt the detached perspective which makes our sociological
observations possible. An illiterate in one of Lerner's Middle Eastern
villages plainly could not do it. The answer is print again, because print
affected the human situation not only in the "sociological" sense, but in
a perceptual sense as well. The process is truly systemic: Print affects
conditions *and* perceptions of conditions which have a further effect on
conditions. In the preceding chapter we discussed the difference between
the classical and modern sets of mind. The modern mind, we observed,
can be understood as that of the detached observer. We must now, in
somewhat greater detail, investigate the way in which print sets up the
detached observer situation, for by this effect print sets the conditions of
modern thought: scientific, religious, philosophical, and for our purposes,
above all, political.

We now need to grasp the substance of Marshall McLuhan's powerful
argument.** McLuhan starts from a premise similar to that of Wittgenstein

---

* *The Passing of Traditional Society,* p. 53. (Italics in original.)
** I am very much in the debt of James W. Carey whose article "Harold Adams
Innis and Marshall McLuhan," *Antioch Review,* Spring, 1967, pp. 5–39, provides
a cogent summary of McLuhan's argument.

and the linguists Edward Sapir and Benjamin Lee Whorf, namely, that the structure of reality is presented to individuals through the medium of language. My remarks on perception at the very beginning of this book derive from Wittgenstein's notion that a form of language is a form of life, that is, that when a man acquires a language he acquires also a way of observing, of organizing his experience, a means of distributing emphasis with respect to the real world. It is in this context that the whole paradigm conception of science advanced by Toulmin, Hanson, and Kuhn makes sense, and they would all, I am confident, acknowledge their debt to Wittgenstein. There are, I think, differences in emphasis in this general line of thought. The linguists Whorf and Sapir are somewhat more likely to suggest that the structure of language *determines* the structure of perception than is Wittgenstein who suggests rather more of a reciprocal relationship between the two. Language can be seen as simultaneously reflecting as well as determining the structure of perception and the relationship is thus more complex. It is the "reflecting" side of the relationship that has given rise to the so-called ordinary language school of modern philosophy.

Where Sapir and Whorf attend to the grammar of language and Wittgenstein to the logic of language, McLuhan, true to his literary and artistic background, emphasizes the mode and form in which it is presented. As James W. Carey sees it McLuhan expands the notion of grammar toward, I would suggest, the form of presentation:

McLuhan does not view the grammar of a medium in terms of the formal properties of language, the parts of speech or morphemes, normally utilized in such an analysis. Instead, he argues that the grammar of a medium derives from the particular mixture of the senses that an individual characteristically uses in the utilization of the medium. For example, language—or better, speech—is the first of the mass media. It is a device for externalizing thought and for fixing and sharing perceptions. As a means of communication, speech elicits a particular orchestration of the sense. While speech is an oral phenomenon and gives rise to "ear-oriented cultures" (cultures in which people more easily believe what they hear than what they see), oral communication synthesizes or brings into play other sensual faculties. For example, in conversation men are aware not only of the sound of the words but also of the visual properties of the speaker and the setting of the tactile qualities of various elements of the setting, and even certain olfactory properties of the person and the situation. These various faculties constitute parallel and simultaneous modes of communication, and thus McLuhan concludes that oral cultures synthesize these various modalities, elicit them all or bring them all into play in a situation utilizing all the sensory apparatus of the person. Oral cultures, then, involve the simultaneous interplay of sight, sound, touch, and smell and thus produce, in McLuhan's view, a depth of involvement in life as the principal communications medium—oral speech—simultaneously activates all the sensory faculties through which men acquire knowledge and share feeling.

However, speech is not the only mass medium, nor must it necessarily be the dominant mass medium. In technologically advanced societies, print, broadcasting, and film can replace speech as the dominant mode through which knowledge and feeling are communicated. In such societies speech does not disappear, but it assumes the characteristics of the dominant medium. For example, in literate communities oral traditions disappear and the content of spoken communication is the written tradition. Speech no longer follows its own laws. Rather it is governed by the laws of the written tradition. This means not only that the "content" of speech is what has previously been written but that the cadence and imagery of everyday speech is the cadence and imagery of writing. In literate communities, men have difficulty believing that the rich, muscular, graphic, almost multidimensional speech of Oscar Lewis' illiterate Mexican peasants was produced by such "culturally deprived" persons. But for McLuhan speech as an oral tradition, simultaneously utilizing many modes of communication, is almost exclusively the province of the illiterate.*

McLuhan senses the relevance of the biological in his analysis and he, thus, serves as yet another example of the convergence of investigation on reality. Here a philosopher, there a social theorist, a historian of science, or psychologist, and in this case a literary analyst get an intimation of the overriding significance of biological and cultural evolution and push their own problems toward solution in its terms. Carey continues:

McLuhan starts from the biological availability of parallel modes for the production and reception of messages. These modes—sight, touch, sound, and smell—do not exist independently but are interdependent with one another. Thus, to alter the capacity of one of the modes changes the total relations among the senses and thus alters the way in which individuals organize experience and fix perception. All this is clear enough. To remove one sense from a person leads frequently to the strengthening of the discriminatory powers of the other senses and thus to a rearrangement of not only the senses but of the kind of experience a person has. Blindness leads to an increasing reliance on and increasing power of smell and touch as well as hearing as modes of awareness. Loss of hearing particularly increases one's reliance on sight. But, McLuhan argues, the ratios between the senses and the power of the senses is affected by more than physical impairment or, to use his term, amputation. Media of communication also lead to the amputation of the senses. Media of communication also encourage the over-reliance on one sense faculty to the impairment or disuse of others. And thus, media of communication impart to persons a particular way of organizing experience and a particular way of knowing and understanding the world in which they travel.

* "Harold Adams Innis and Marshall McLuhan," p. 16.

Modes of communication, including speech, are, then, devices for fixing perception and organizing experience. Print, by its technological nature, has built into it a grammar for organizing experience, and its grammar is found in the particular ratio of sensory qualities it elicits in its users. All communications media are, therefore, extensions of man, or, better, are extensions of some mix of the sensory capacities of man. Speech is such an extension and thus the first mass medium. As an extension of man, it casts individuals in a unique, symbiotic relation to the dominant mode of communication in a culture. This symbiosis is not restricted to speech but extends to whatever medium of communication dominates a culture. This extension is by way of projecting certain sensory capacities of the individual. As I have mentioned, speech involves an extension and development of all the senses. Other media, however, are more partial in their appeal to the senses. The exploitation of a particular communications technology fixes particular sensory relations in members of society. By fixing such a relation, it determines a society's world view; that is, it stipulates a characteristic way of organizing experience. It thus determines the forms of knowledge, the structure of perception, and the sensory equipment attuned to absorb reality.

Media of communication, consequently, are vast social metaphors that not only transmit information but determine what is knowledge; that not only orient us to the world but tell us what kind of world exists; that not only excite and delight our sense but, by altering the ratio of sensory equipment that we use, actually change our character.*

Print in addition to the multitude of "sociological" effects which we suggested a moment ago has an even more profound effect. It alters the conditions of perception because it isolates the visual sense and makes it dominant.

Print, the dominant means of communication in the West, depends on phonetic writing. Phonetic writing translated the oral into the visual; that is, it took sounds and translated them into visual symbols. Printing enormously extended and speeded up this process of translation, turning societies historically dependent upon the ear as the principal source of knowledge into societies dependent upon the eye. Print cultures are cultures in which seeing is believing, in which oral traditions are translated into written form, in which men have difficulty believing or remembering oral speech—names, stories, legends—unless they first see it written. In short, in print cultures knowledge is acquired and experience is confirmed by sight: as they say, by seeing it in writing. Men confirm their impressions of Saturday's football game by reading about it in Sunday morning's paper.

Besides making us dependent on the eye, printing imposes a particular logic on the organization of visual experience. Print organizes reality into discrete, uniform, harmonious, causal relations. The visual arrangement of the printed page becomes a perceptual model by which all reality is organized.

* "Harold Adams Innis and Marshall McLuhan," pp. 17–18.

The mental set of print—the desire to break things down into elementary units (words), the tendency to see reality in discrete units, to find causal relations and linear serial order (left to right arrangement of the page), to find orderly structure in nature (the orderly geometry of the printed page)—is transferred to all other social activities. Thus, science and government, art and architecture, work and education become organized in terms of the implicit assumption built into the dominant medium of communication.

Moreover, print encourages individualism and specialization. To live in an oral culture, one acquires knowledge only in contact with other people, in terms of communal activities. Printing, however, allows individuals to withdraw, to contemplate and meditate outside of communal activities. Print thus encourages privatization, the lonely scholar, and the development of private, individual points of view.

McLuhan thus concludes that printing detribalizes man. It removes him from the necessity of participating in a tightly knit oral culture. In a notion apparently taken from T. S. Eliot, McLuhan contends that print disassociates the senses, separating sight from sound; encourages a private and withdrawn existence; and supports the growth of specialization.

Above all, print leads to nationalism, for it allows for the visual apprehension of the mother tongue and through maps a visual apprehension of the nation. Printing allows the vernacular to be standardized and the mother tongue to be universalized through education.

While the book ushered in the age of print, developments such as newspapers and magazines have only intensified the implications of print; extreme visual nationalism, specialist technology and occupations, individualism and private points of view.*

When print emerges as the dominant mode of communication in the West, it, because of the isolation of the visual sense and of the isolation of the reader, emphasizes and makes dominant the position of the detached observer. There are two important points to be made here. The first is at the sociological level. The written word in manuscript culture, as we have previously noticed, carried with it a pervasive intellectual authority. Print breaks the monopoly on knowledge; slowly at first, then in an escalating fashion. But the power of the written word, now in printed form, continues. When writing is freed from parchment, men and their minds are freed as well. The detached observer of nature, psychologically liberated from the authority of the ancient texts, begins to experiment with the manipulation of words as a way to unlock the secrets of nature. Remember our earlier discussion of Galileo's experiments with the contrary to fact notion of bodies falling in a vacuum. Thomas Hobbes, the modern political theorist *par excellence,* vehemently discards the tradition of the ancient texts and, significantly, demands that men fashion political *words*—"sovereignty," for example—anew. And these are *English* words he talks

* "Harold Adams Innis and Marshall McLuhan," pp. 18–20.

about. The "state" shall become what men make it, and not what the ancients have decreed. Machiavelli ignores the tradition of natural law and writes *in Italian* of politics as he sees it. Luther overthrows the tradition of the church, *translates the Bible into German,* and demands a priesthood of all believers. I would argue here no simple cause and effect relationship, but rather the notion that print made it possible for these efforts to succeed and in some measure conditioned the attempts. And we cannot ignore the strides made in natural science through and because of inventions in mathematical symbolism. This is crucial to our scientific culture. Newton, for example, could do what he did because he felt himself free to invent a new calculus.

We are led, thus, to the second point which relates to McLuhan's observation that print isolates the visual sense. Men came to understand the problem of knowledge as a matter of relating the visual, printed word, symbol, or proposition to reality. On the one hand this way of thinking led to an uncertainty with respect to the nature of reality—if one understood the world through written symbols their accuracy became a matter open to question. On the other the notion quickly developed that man could manipulate the world by manipulating the words and symbols used to describe it. The creative manipulation of words and symbols is what modern science and philosophy are all about. The recognition of the limitations on the accuracy of words and symbols is what empiricism, a distinctly modern philosophical perspective, is all about. All of this is captured in the image of a man with a book in hand. He looks at the printed phrase and looks out at the world there described. If the words fit, he is satisfied. If they do not fit, he seeks to fashion new words, phrases, or symbols that do fit. The words are authoritative, but they are not sacred.

Ludwig Wittgenstein, the contemporary philosophers like to say, was twice a genius. In his later life he fashioned in his *Philosophical Investigations* the foundation of the ordinary language school of philosophizing. To this latter teaching we have already made numerous references and we shall say more in the next chapter. Let us for the moment, however, look at his first act of genius, the writing in the years of World War I of the *Tractatus Logico-Philosophicus.* We are now, I think, prepared to see clearly what it was that he did. The isolation of the visual sense was introduced into the Western world with print. It worked its ways on the Western mind for four hundred years changing, primarily through its product, modern science, the condition of human kind. Philosophers grasped for an understanding of its impact. Locke, Hume, and Kant in particular reflected its power and flirted with stating its nature. But they were all involved in the mainstream of the Western philosophical tradition and their insights were mixed with the *content* of the writings of the past— with, if you will, the spell of Plato and Aristotle.

The young Wittgenstein came to logic and epistemology not through the academic philosophical tradition but through first engineering and then the foundations of mathematics. He confronted print culture thinking head on; he distilled its essence and wrote it down for all to see. The Wittgenstein of the *Tractatus* is generally (whether justly or not) said to have been the founder of Logical Positivism, and Logical Positivism is understood to be the ultimate statement of scientific empiricism. The core of his doctrine is what has come to be called the picture theory of meaning, Wittgenstein's way of showing *how* a proposition means, of how language describes reality. Wittgenstein's notion is that a proposition has meaning because it pictures a possible state of affairs in the world. The academic philosophers who followed him under the name Logical Positivists tended to be nervous about this doctrine and tended to shy away from it, but Wittgenstein was utterly serious about it. The following is the English translation of the heart of his statement and, unusual and difficult as it is, I must ask you to read it carefully. The numbers are Wittgenstein's way of ordering the statements in his argument.

2.1	We make to ourselves pictures of facts.
2.11	The picture presents the facts in logical space, the existence and non-existence of atomic facts.
2.12	The picture is a model of reality.
2.13	To the objects correspond in the picture the elements of the picture.
2.131	The elements of the picture stand, in the picture, for the objects.
2.14	The picture consists in the fact that its elements are combined with one another in a definite way.
2.141	The picture is a fact.
2.15	That time elements of the picture are combined with one another in a definite way, represents that the things are so combined with one another. This connexion of the elements of the picture is called its structure, and the possibility of this structure is called the form of representation of the picture.
2:151	The form of representation is the possibility that the things are combined with one another as are the elements of the picture.

2.1511   Thus the picture is linked with reality; it reaches up to it.

2.1512   It is like a scale applied to reality.

2.15121   Only the outermost points of the dividing lines *touch* the object to be measured.

2.1513   According to this view the representing relation which makes it a picture, also belongs to the picture.

2.1514   The representing relation consists of the coordinations of the elements of the picture and the things.

2.1515   These coordinations are as it were the feelers of its elements with which the picture touches reality.

2.16   In order to be a picture a fact must have something in common with what it pictures.

2.161   In the picture and the pictured there must be something identical in order that the one can be a picture of the other at all.

2.17   What the picture must have in common with reality in order to be able to represent it after its manner—rightly or falsely—is its form of representation.

And further:

4.01   The proposition is a picture of reality.
The proposition is a model of the reality as we think it is.

4.011   At the first glance the proposition—say as it stands printed on paper—does not seem to be a picture of the reality of which it treats. But nor does the musical score appear at first sight to be a picture of a musical piece; nor does our phonetic spelling (letters) seem to be a picture of our spoken language. And yet these symbolisms prove to be pictures—even in the ordinary sense of the word—of what they represent.

4.012   It is obvious that we perceive a proposition of the form *aRb* as a picture. Here the sign is obviously a likeness of the signified.

4.013   And if we penetrate to the essence of this pictorial nature we see that this is not disturbed by *apparent irregularities* (like the use of ♯ and ♭ in the score). . . .

4.021   The proposition is a picture of reality, for I know the state of affairs presented by it, if I understand the proposition. And I understand the proposition, without its sense having been explained to me.

4.022   The proposition *shows* its sense.

The proposition *shows* how things stand, *if* it is true. And it *says,* that they do so stand.

4.023   The proposition determines reality to this extent, that one only needs to say "Yes" or "No" to it to make it agree with reality.

Reality must therefore be completely described by the proposition.

A proposition is the description of a fact.

As the description of an object describes it by its external properties so propositions describe reality by its internal properties.

The proposition constructs a world with the help of a logical scaffolding, and therefore one can actually see in the proposition all the logical features possessed by reality *if* it is true. One can *draw conclusions* from a false proposition.

4.024   To understand a proposition means to know what is the case, if it is true.

(One can therefore understand it without knowing whether it is true or not.)

One understands it if one understands its constituent parts.*

* Ludwig Wittgenstein, *Tractatus Logico-Philosophicus* (New York: Humanities Press and London: Routledge & Kegan Paul Ltd., 1922), pp. 39–41, 63–65, 67.

"Die Welt ist alles, was der Fall ist" ["The world is everything that is the case,"]. Thus Wittgenstein begins his *Tractatus*. The set of genuine propositions pictures what is the case or what might conceivably be the case. Ordinary parlance may conceal this correspondence, but the logically pure language would picture it. Wittgenstein's dominant image is of the written or printed phrase as his emphasis on the picture shows. Note especially his simile "like a scale applied to reality." It was, I think, the fact that he saw the implications of the logic of print so clearly that he found it necessary to reject his doctrine of the *Tractatus* and to turn later in his life to the logic of use, the logic of the living, spoken language. Let me leave these considerations for the moment, and return more directly to political artifice in the modern West.

I wish to argue not that the logic of print completely and utterly replaced the logic of the spoken word and the manuscript in Western culture. The logic of print became most completely dominant in the realm of natural science and it remained weakest in the realm of religion and theology. It is this fact that makes our earlier contrast between the church service and the logic class a helpful illustration. It is a commonplace, but an accurate one, to understand modern Western culture as bifurcated, as a battle between religion and science. There has been and continues to be an obvious struggle at the institutional level. The typical Western Christian fights out the problem in his individual consciousness in a way described by countless writers ranging from Turgenev to Camus.* Philosophy and especially political theory since 1500 lies between the poles of science and religion and the struggle for accommodation is the underlying theme of all of it. We can begin to see the issues involved by investigating in a general way the development of modern moral and political thought.

10

A discussion of moral reflection after the scientific revolution which pretended to anything like thoroughness would require several volumes and I can make no more than some general observations about it here.** We are accustomed to regarding the development of modern science as a positive force for human emancipation of enormous proportions. Man at long last came to see the universe as knowable, we are inclined to say,

* See Turgenev's *Fathers and Children* and Camus's *The Stranger* and *The Myth of Sisyphus*.
** Harry Prosch, *The Genesis of Twentieth Century Philosophy* (New York: Doubleday & Company, 1964) is as good a general discussion of the philosophical developments as I know of.

and he threw off the shackles of tradition and superstition and proceeded to bring the universe under his intellectual and practical control. In a very important sense this judgment is quite accurate. From the perspective of moral thought, however, the situation looks substantially different. What formerly was obvious, for example, the existence of the human soul constructed in the image of a personal God, becomes quite doubtful. Questions which could formerly be answered confidently and rather easily, say, the proper purpose of human life in the hands of Aristotle or Thomas Aquinas, became enormously complicated and perhaps in the final analysis intractable.

Science made the universe seem knowable in a new and wonderful way as long as the knower himself was totally removed from the object of knowledge. As long as the personal consciousness of the investigator was wholly detached he could in principle come to understand all of the objects and processes outside himself, that is, his personal consciousness, including (when behavioral psychology and Freudian psychological theory came on the scene) his own behavior. Man, provided with a detached point of view by print, finally achieved his Archimedean point. This point of detachment, from which as Archimedes had prophetically suggested man could lift the world, resulted in the establishment for Western man of two new and important philosophical perspectives. One is credo: the demand for detached objectivity which we have come to call scientific method. The other is a point of departure: the absolute separation between mind, that is, personal consciousness, and body.

When the human mind (as opposed to the human brain) found itself outside natural processes looking in, enormous possibilities for the creative gathering of new knowledge were opened and these possibilities, speaking relative to man's time on earth, were very rapidly realized. But at the same time this position of detachment tended to make all creations of the mind which could not be tested against natural processes not objective as they had previously been but radically subjective. Speaking now in terms of a much shorter time scale, from 1500 to the present, this condition of subjectivity did not happen all at once. The seeds, however, were sown early and the logical conclusion was perhaps first clearly stated by David Hume in the eighteenth century although the ultimate statement surely belongs to the Logical Positivists in the twentieth century. For a Logical Positivist like A. J. Ayer the products of the human mind were divided into two categories: sense and nonsense. The sense category included only empirical propositions which could be tested against sense experience, and definitional systems like mathematics. All other mental emissions whether ethical, esthetic, poetic, or metaphysical were literally nonsense or, to put it another way, in the final analysis merely subjective. The characteristic premise of existentialist thought is the personal awareness of the existence of personal consciousness, which is but a different perspective on the situation which called Ayer's analysis into being.

Now it is perfectly clear that all moral reflection in the West since 1500 has not been Logical Positivist or existentialist. What I would suggest, however, is that all moral reflection in the West since the development of modern science has sought to deal with the problem stated by Logical Positivism in insoluble form. The problem for moral thought has, thus, been one of accommodation, repair, or, as John Dewey once put it, reconstruction. I think that I can state with only the most minor of qualifications that no modern moral thinker whether he be philosopher, novelist, poet, dramatist, or theologian has been able to ignore the problem created by the breakdown of the classical *Weltanschauung*. Approaches to this problem have varied all the way from reveling in subjective emotion on the one side to the complete objectification of human consciousness on the other. The approaches are so various that I cannot catalogue them all. I would like, however, to describe some of the most important. Starting in each case from the situation of detachment described earlier, I shall try in the succeeding paragraphs to state the crucial steps in several representative and important attempts at solution.

(1) If it is true that human consciousness and choice are detached from the natural order of things, it should be possible for man after understanding the relevant features of the natural situation to create by his own choice an effective moral order. The natural fact is that men are selfish, security-oriented animals living in constant fear of death. Thus, they must create *de novo* an order which so limits each man's aggressiveness that all are secure. The proper political and moral order is, thus, not to be found in nature but literally to be created and invented by man himself. This is the view of Hobbes and in large measure the view of the framers of the American Constitution.

(2) The human consciousness is detached from the natural order to be sure, but its incarceration by rules, even those created by men (as recommended by Hobbes), will frustrate its creativity. The invention of rules is an attempt to impose order where it does not belong. Value is to be found *in* the subjectivity, not in shackling it. If art and morality are emotion, then let them be emotion and not rule-construction. This, of course, is the Romantic view which perhaps ultimately found its flowering in the notion of the folk-soul.

(3) The principal receptacle of the classical view of man was and is the Roman Catholic Church. For Catholic thinkers the solution lay either in denying the situation of detachment by contending that natural science was not what it seemed to be, arguing that the order of nature contained two separate realms (the human and the physical) operating according

to different principles or, most usually, in some combination of both contentions.

(4) The remaining possibility is to deny that human consciousness is what it seems to be. Human consciousness is not really detached from natural processes, it is only a particularly complex form of natural process. This objectification of consciousness characterizes nineteenth-century social science positivists like Comte and Pareto, psychologists like Watson and in some measure Freud himself, and is implicit in a good deal of contemporary social science. Of course, the most significant representative of this view is Kárl Marx who suggested that consciousness was but a reflection of the interaction at a particular time of man as physical object with other physical objects. Thus, moral systems were "residues" for Pareto, rationalizations or Oedipal products for psychologists, and superstructure for Marx.

Our culture can reasonably be characterized by the all-pervasiveness of modern science and modern scientific standards. This is what makes our modern problem a problem of scientific culture in the broadest sense and not simply one of technological change. Our very common sense is informed and molded not only by what science has substantively taught us about the world we live in, but also by the criteria which determine what will count as understanding a phenomenon or process. Every field of intellectual endeavor in the modern West has to come to grips with scientific standards, sometimes dressed in the resplendent uniform of Scientific Method and sometimes under the guise of common sense. For this reason speaking of intellectual activity in terms of separate and wholly self-contained realms, that is, as if there were moral criteria, esthetic criteria, as well as scientific criteria, is often not very helpful. It is easy to assume that the moralist has criteria which are entirely unrelated to science. Unfortunately, however, it is not so. The serious commentator on ethics in the modern world must, as our four prototypes above illustrate, either operate in terms of scientific standards or go to great lengths to explain them away. His argument would be hollow and irrelevant to the modern situation if he pretended that science and standards of scientific explanation did not exist.

The problem for contemporary moral thought is nicely stated by Leo Strauss in terms of the question of justifying natural rights:

> The issue of natural right presents itself today as a matter of party allegiance. Looking around us, we see two hostile camps, heavily fortified and strictly guarded. One is occupied by the liberals of various descriptions, the other by the Catholic and non-Catholic disciples of Thomas Aquinas. But both armies and, in addition, those who prefer to sit on the fences or hide their heads in the sand are, to heap metaphor on metaphor, in the same boat. They are all modern men. We are all in the grip of the same difficulty.

Natural right in its classic form is connected with a teleological view of the universe. All natural beings have a natural end, natural destiny, which determines what kind of operation is good for them. In the case of man, reason is required for discerning these operations; reason determines what is by nature right with ultimate regard to man's natural end. The teleological view of the universe, of which the teleological view of man forms a part, would seem to have been destroyed by modern natural science. From the point of view of Aristotle—and who could dare to claim to be a better judge in this matter than Aristotle?—the issue between the mechanical and the teleological conception of the universe is decided by the manner in which the problem of the heavens, the heavenly bodies, and their motion is solved. Now in this aspect, which from Aristotle's own point of view was the decisive one, the issue seems to have been decided in favor of the nonteleological conception of the universe. Two opposite conclusions could be drawn from this momentous decision. According to one, the nonteleological conception of the universe must be followed up by a nonteleological conception of human life. But this "naturalistic" solution is exposed to grave difficulties; it seems to be impossible to give an adequate account of human ends by conceiving of them merely as posited by desires or impulses. Therefore, the alternative solution has prevailed. This means that people were forced to accept a fundamental, typically modern, dualism of a nonteleological natural science and a teleological science of man. This is the position which the modern followers of Thomas Aquinas, among others, are forced to take, a position which presupposes a break with the comprehensive view of Aristotle as well as that of Thomas Aquinas himself. The fundamental dilemma, in whose grip we are, is caused by the victory of modern natural science. An adequate solution to the problem of natural right cannot be found before this basic problem has been solved.*

The problem, in short, is one of getting moral concerns into a proper relationship with the natural order, of getting moral philosophy properly adjusted to standards of scientific understanding or alternatively of getting standards of scientific understanding properly adjusted to moral philosophy. The four prototypes discussed earlier all in some sense were attempting to do this. Only a dogmatic adherent of one of the views would be inclined to regard his particular notion as simply correct and the others as wholly wrong. Each view evaluated separately seems at least plausible, but clearly all cannot be wholly correct. A pattern in the evolution of ideas shows itself. The problem is set by the breakdown of the universal view characteristic of earlier times—the printing press is to be found right at the center of this breakdown—and various attempts are made to solve the problem in terms of the tools available after the breakdown.

Like a group of overlapping circles proposals are advanced which

* Leo Strauss, *Natural Right and History,* Chicago: University of Chicago Press, 1953. Copyright University of Chicago Press, 1953, pp. 7–8.

strive to make sense of political problems in terms of a particular focus. This pattern is observable at a variety of levels. The focus may be rule creation as with Hobbes or relationship to physical objects as with Marx; or successive and competing foci can be detected at less grand a level. I am particularly intrigued by the analysis of the concept of representation lately presented by Hanna Pitkin.* Mrs. Pitkin describes in detail the development of the concept beginning with Hobbes. In so doing she presents a precise picture of the sort of "overlapping circle" progression of concepts that I am attempting to describe. The "authorization" conception of representation overlaps with the "accountability" conception—they are part of a developing family of meanings. Equally intriguing in terms of our discussion is Mrs. Pitkin's Appendix on Etymology. The Greeks it seems had no word for representation in the modern sense. While the term is of Latin origin, "its original meaning had nothing to do with agency of government or any of the institutions of Roman life which we might consider instances of representation." ** In the Middle Ages the term was used to describe the fact that the church leaders are the embodiment of Christ and the Apostles, but not, significantly, their representatives in the modern sense. The "stand for" sense of representation did not emerge until roughly the fifteenth century. One cannot read Mrs. Pitkin's account without being struck by the fact that the meaning of this significant word took on the logic of visual detachment—something which it had never had before—in the fifteenth and sixteenth centuries. And Hobbes is significant for Mrs. Pitkin's analysis because he, in the 1640s and 1650s, was the first to make anything of the notion that the government "stood for" or "represented" the people.

What we are talking about here are focal ideas, what in the context of the history of science Thomas Kuhn has called paradigms. Sheldon Wolin sees the parallel:

> When applied to the history of political theory, Kuhn's notion of a paradigm, "universally recognized scientific achievements that for a time provide model problems and solutions to a community of practitioners," invites us to consider Plato, Aristotle, Machiavelli, Hobbes, Locke, and Marx as the counterparts in political theory to Galileo, Harvey, Newton, Laplace, Faraday, and Einstein. Each of the writers in the first group inspired a new way of looking at the political world; in each case their theories proposed a new definition of what was significant for understanding that world; each specified distinctive methods for inquiry; and each of the theories contained an explicit or implicit statement of what should count as an answer to certain basic questions. Kuhn's criterion, that a paradigm should provide "model problems and solutions," is approximated in the way that one theorist will

---

* Hanna Fenichel Pitkin, *The Concept of Representation* (Berkeley and Los Angeles: University of California, 1967).
** *The Concept of Representation,* p. 241.

adopt major elements from another. When Thomas Aquinas refers to Aristotle as "the philosopher" and proceeds to incorporate certain key Aristotelian notions, such as *physis* and *polis,* and put them to work, we have a striking analogy with what Kuhn calls "paradigm-adoption." Many other instances could be introduced to show that the tradition of political theory displays a high degree of self-consciousness about the role and function of paradigms. Harrington's political ideas were elaborated in reference to two main paradigms, that of "ancient prudence" as represented by Aristotle and that of "modern prudence" by Machiavelli. One could also point to the paradigmatic influence of Locke upon the eighteenth-century political writers in America and France; of Hobbes upon later writers such as Bentham, James Mill, and Austin; of Marx and Max Weber upon political and social writers of the last hundred years.*

But, of course, political theory presents paradigms of political understanding not only to theorists but, in the modern world at least, to the general public as well. The mass media, led by the printing press, create an entirely new political situation. The paradigm must be sold to large numbers of people and political theory slides into what we have come to call ideology. Much has already been said about this subject and I have no reason to repeat it here. What I do want to stress is that the "overlapping circle" image, the competition of paradigms or focal ideas is also visible when one looks at the great ideologies.

The problem of modernism is created by the emergence of the commercial, industrial, secular, mass-oriented nation-state out of the ruins of the traditional community stabilized by ancient authority. And, as we have said, the printing press lies right at the heart of this development, because printing not only involved the masses in activities which had formerly been none of their business, it was also the first mass-production industry. The great modern ideologies—liberalism, nationalism, and communism—all grapple with the problem of modernism according to a particular focus. The individual, the national group, and the class are respectively made the focus of doctrines which attempt to resolve the conflict between the human and the technical. These doctrines can harden into dogma as the experience of the Germans and the Russians has surely shown and perhaps the experience of the Americans as well. It is against this background, against the logic of evolutionary experience, that a political orientation for the latter half of the twentieth century must be hammered out.

* Preston King and B. C. Parekh, editors, *Politics and Experience* (Cambridge at the University Press, 1968) "Paradigms and Political Theories," Sheldon S. Wolin, pp. 140–141. It is not surprising, I think, that Wolin should see this parallel, for his *Politics and Vision* anticipates the idea of evolution through paradigm. Indeed, I would urge anyone who finds my attempt of the last two chapters either interesting or annoying to read Wolin's excellent central chapters on Luther, Calvin, Machiavelli, and Hobbes.

10
Evolution Become Conscious of Itself

1

In the preceding chapters we have struggled with the task of establishing a new perspective on politics. We have been bold where we might have been cautious. Some may indeed suggest that we have been reckless, sweeping an extraordinary diversity of factors into what may seem a simple mold. In an important sense, however, I would argue that we have done nothing very unusual. I have only sought to state explicitly what is in fact all around us. Sir Julian Huxley, in a phrase which appealed to Teilhard de Chardin, has described the contemporary intellectual condition of man as evolution become conscious of itself. If one argues as Huxley and Teilhard do that human consciousness is a product of the evolutionary process, implicit in it from the very beginning, then man since Lyell and Darwin has begun to become evolution conscious of itself.

The process is by no means complete. All men have not yet realized the character of their connection with nature and with the past, but the recognition of biological reality bubbles out of the centers of Western culture in myriad ways. As I have argued, the philosophers and historians of science have seen an aspect of this reality, although this is not to say that all would recognize or admit that this is what they have seen. Freud saw that each man begins as a child and that men began as animals called primitive men. The enormous impact of psychoanalysis on our life and literature follows from this insight. The anthropologists saw evolution, rejected it temporarily, and are now rediscovering it in a sophisticated way. The ethologists have just begun to supply us with details in a genuine natural science of behavior. Innis, McLuhan, and communications theorists have helped to open our understanding of the mechanism of cultural evolution—although they have not understood their own work in quite this way. Economists, political scientists, and sociologists, whether they like it or not, have been forced to face the fact of development. And Père Teilhard, from his unique perspective, saw the whole thing, even though he supplies us with little of a specific nature with respect to cultural evolution.

It is against this background that I have tried to make sense of human organization, not, to be sure, in its details but in a broad perspective in terms of which the details can be intelligible. The level of human understanding and technique, which we have come to call science and technology, brought to us by a culture dominated by print, has opened up the understanding in time and evolution. The world can never be the same again. Père Teilhard with his customary acuteness stresses the importance of the spherical nature of the world. The world is round, not flat or parabolic, and culture therefore spreads around the globe and humanity turns it on itself. The electronic media, as McLuhan suggests, pick up where print leaves off and we march toward the global village. Teilhard's noosphere, the ocean of ideas and artifacts that surrounds the globe, is a reality in a way that it was not a hundred or two hundred years ago.

We have only to look around to see the overriding political significance of the particular way in which what we call modern culture has come to the world. Contemporary Western society rocks and trembles over the bifurcation of life into technical and human spheres. What Marx saw from his limited nineteenth-century point of view as the alienation of man from himself by the machine, other observers have seen in accordance with other dichotomies: church and secular society, community and abstractly defined individual roles, literature and science, personal security of family and insecurity in the rat-race, and in perhaps the most extreme statement the problem of personal existence in an absurd world. We hear of the politics of mass society, of anomie and ideology, of the escape from freedom. We have seen the Germans try to combine a technical war machine with a return to tribalism and the Russians attempt to industrialize through enforced community. And today we witness political leaders from Mao Tse Tung to Julius Nyerere struggle with the problem of grasping the fruits of technical modernity without losing the spiritual balance central to the ancient manuscript and oral tradition. For Mao's *cultural* revolution does battle with the bureaucratic accouterments of technological advance as exemplified by the "revisionist" Russians and Nyerere's African Socialism seeks to preserve the spirit of tribal community. The print-culture American A.I.D. official sees his dollars "corruptly" swallowed up by a family oriented morality that is a fundamental part of oral culture society and can understand it only as nepotism.

What I suggest by all this is that the evolutionary paradigm of understanding as I have attempted to sketch it here points up problems for research that the universal-generalization model simply cannot see. The important question becomes not in what way is the same function, say, political socialization, performed in all societies, but what effects does essentially oral socialization have in contrast with print socialization? Recognizing the importance of artifice, of human choice, in political evolution, researchers must come to ask with greater seriousness: What do the choices with respect to political and economic development look like to the chooser? Contemporary politics in various parts of the world is marked by the presence of oral culture men who have been print trained and who must make choices with respect to "modernizing" their ancient societies.

The process is not automatic; it is not entirely predictable, but it can be *understood* against a background of cultural evolution. What I am saying, in short, is simply that taking time and evolution seriously can open the door to a genuine science of society, one which is capable of grasping the whole phenomenon of man. This is science not because of what it can be made to look like by the clever mimic, but because it is fundamentally compatible with—indeed a natural and logical extension of —biology, chemistry, and physics. The genetic system of information transmission is ultimately a matter of physics and chemistry. The cultural system of information transmission is different in many ways, but in the final analysis it is tied through evolution to the same natural processes. Whatever deserves the name political science must start with a conception of the human animal in nature and must conclude with an account which will deal with the whole phenomenon of politics, the choosing and creating as well as the reacting and responding.

2

The classical world, the world of Plato and Aristotle, saw man in nature but it could not see man in time. The Christian world of Augustine and Thomas Aquinas saw a dichotomy between man's spiritual life and his temporal life, between soul and body, and it saw man in time. A fore-shortened but nonetheless linear conception of time is clearly present in Christian thinking—creation, fall, immaculate conception, crucifixion, resurrection, second coming—and the separation of body and soul presages the notion of man's detachment from nature. The modern world —in the historians' sense of modern, say from 1500 A.D. to 1800 A.D.— kept the truncated Christian conception of time but saw man as detached from nature, as artificer *par excellence.*

We have touched on this history before and I summarize it here for a particular purpose. I want now to turn our discussion toward political philosophy, toward general political recommendation. In so doing I want insofar as possible to avoid being misunderstood. The notion that I hope to get across is that the intellectual situation for political philosophy, and for that matter for philosophy in general, during the last 100 to 150 years has been utterly unique. I stretch the period to 150 years because, retrospectively, one can see the trends forming that long ago. In the important sense, however, this utterly new philosophical perspective could only be fully grasped by twentieth-century man. Students of political philosophy in the twentieth century have, however, almost entirely failed to understand this fact and have argued for some sort of restoration of the classics. Likewise, they are inclined to reduce whatever they find appealing in contemporary writings to identity with the teachings of a classical writer. Thus, what we have said here might be understood simply as a sort of recapturing of the man-in-nature perspective of Aristotle. There is, of course, truth in this, but I should like to stress the important difference.

Only in the last hundred years or so could any man have possibly known (1) that man is the product of an evolutionary process stretching back some six billion years, and (2) that man was capable of establishing the detached observer position, of devising modern science and technology, and of creating modern society. Plato, Aristotle, Augustine, and Thomas Aquinas—masters though they were—could not possibly have known of these parameters. Descartes, Hobbes, and Locke could see the position of detachment and some of its implications, but by the same token man's connection with the rest of nature had to be rejected and the significance of time could not have been grasped. They were, moreover, too busy creating modern science and modern society to step back and reflect on the implications of the possibility.

Twentieth-century man, however, can see the vastness of time and its overriding significance, man's connection with the natural world through evolution, and the fact of modern culture, its intellectual style and its material product. The recognition of these factors—all of them together— provides a standard, a perspective, utterly unique in human history in terms of which man can decide what he *ought* to do. In this sense a new standard for philosophy and for political philosophy in particular becomes possible. What arises from this perspective is a new understanding of recommendation and justification, a new logic of recommendation, which can only make full sense in the twentieth century.

In a curious and in some measure incomplete way the first thinker to grasp the implications of this contemporary perspective was Nietzsche. This is not the place for a full dress discussion of Nietzsche's ideas, but I do need to briefly describe the core of his teaching so that we can get into focus what might be described as his logical discovery. Animals, Nietzsche argues, forget each moment as soon as it passes. Living, thus, completely in the present means to live unhistorically. Man on the contrary is the animal that cannot help remembering the past, and he therefore lives historically. He does not, however, remember everything. To do so would in fact condemn him to perpetual agony. Only by selectively remembering, being able to forget, can man be happy. Nietzsche, anticipating some of the things that we have said here about the connection between biological and cultural evolution, suggests that this capacity to remember, to turn the experiences of the past into tools for the present, is what separates man from other animals. The dividing line between what an animal, including man, knows, that is, what he remembers, and what he does not know, Nietzsche calls his horizon. Men must have horizons but different men have different horizons—horizons vary over time and with experience. No horizon is or can be completely true; thus, all horizons are false or, in sum, everything is false. These are the conclusions that Nietzsche arrives at by applying critical scientific standards to the human animal developing through time.

All men have lived in accordance with horizons, but all horizons are false—they do not and cannot describe reality as it is. Now the uniquely contemporary twist: This recognition points not to despair but to man's creativity and his power. Man can now, and only now, be seen as the animal capable of creating horizons. For the first time in history he can consciously create his own horizon. Might it be, Nietzsche suggests, that the horizon creator, conscious of himself as horizon creator, will create the most glorious, most genuinely human horizon of all?

I do not wish to argue the merits or demerits of Nietzsche's particular way of stating the point in these pages. I ask you only to attend to the logic of his way of arriving at normative conclusions. His assertion, aphoristic and perhaps excessively dramatic as it is, that the search for truth about reality can never end, that, therefore, any and all horizons with respect to reality are false, leads to and justifies a directive. The directive tells not what is the case but what to do, and in this respect is normative in the full sense of the word. The guide for conduct consists in the recognition that all previous guides, because they rest upon a false conception of reality, are false.

The key to this curious argument is that Nietzsche saw clearly a tension between truth and life. The normative, the fact that man must *do* something and he must decide *what* to do, is a matter of life, of the biological fact of existence. Truth in the sense of the accurate description of reality can never be obtained—all attempts are false—but life is the paramount matter; it proceeds, truth or no truth. When Albert Camus says half a century later that suicide is the only important philosophical question— one must first decide whether life should be lived or not—he is following Nietzsche. Nietzsche strikes at previous philosophizing which argued in the form "*x* should be done *because* such-and-such *is* the case." Such assertions are always false, but the recognition of this falsity opens the door to the free creation of the norms of life. Life and truth are not identical. Truth is not prior to life. Life is prior to truth.

Nietzsche is supremely conscious of the fact that his philosophical problems are problems of his time and culture—problems unique to nineteenth-century Europe. His famous suggestion that God is dead means that he is dead for nineteenth-century Europeans even though he once lived for Europeans of an earlier time. Nietzsche writes *after* the discovery of the importance of history, *after* the promulgation of man's animal roots in the theory of evolution, and *after* the Western mind had devised the distinction between "is" and "ought." My argument here is that his position would have made no sense—or better, perhaps, would have been impossible—apart from this unique perspective.

We have, I think, said enough to make the first two aspects of his perspective clear. I should, however, say a word or two more about the significance of the separation of "is" and "ought." It is a fact that the is-ought distinction is made the object of heavy emphasis nowhere save in the modern West. This fact, it seems to me, is to be understood in terms of the detached observer position which characterizes the modern West. The modern observer assessing the world visually through the medium of the printed proposition (*a la* Wittgenstein in the *Tractatus*) sees a radical is-ought separation because of his perspective. *Looking* at the world through the lens of the printed proposition makes one *see* that while the "is" is plainly "out there," the "ought" is nowhere to be found (save perhaps in the mind of the observer). Primitive, classical, medieval, or for that matter oriental man because of his sensory engulfment in nature —while he knows, in some sense, that there is a difference between "is" and "ought"—does not feel the radical separation. The message of classical moral philosophy is plainly that by knowing what man *is,* his essence, his nature, one can know what he *ought* to do, what is right or wrong for him. Nietzsche, in the spirit of Hume and the Logical Positivists, sees that knowledge of the essence of things, including man, is illusory— all horizons are false—but his brilliance, largely unnoticed, consists in his positive conclusion from this uniquely modern insight. Saying, with the

Logical Positivists, that ought statements are emotive is to say very little. It is like saying all men are warm-blooded; it provides no intellectual tools for distinguishing one from another or choosing one over another. Nietzsche in his way recognizes ought statements as emotive, as a matter of "will," but by proclaiming the supremacy of life over truth turns the recognition toward positive creativity.

3

If the results of scientific investigation in biology, geology, palaeontology, and related fields show us that the human world is more or less as we have described it here, that is, as a matter of animal adaptation over time by means of a peculiar and complex form of information transmission, we are directed, when we turn to problems of philosophy, to ask a rather different kind of question with respect to philosophers. The question is not so much "What did he mean?" although this is surely not irrelevant, but rather "What would the world have looked like given his particular perspective?" And, as I have argued all along, perspective is not only a matter of what problems were present, but also of what tools—both media of communication and kinds of conceptual apparatus—were available. The remarks previously made about Plato, Aristotle, Hobbes, and for that matter about Nietzsche should be understood as responses to this kind of implicit question.

I want now to say a word or two about the contemporary lines of philosophical inquiry known as existentialism and analytic philosophy. It is perfectly clear from the outset that these two lines of thought cut into the philosophical Neapolitan ice cream pie from very different angles. In the context of the bifurcated Western culture already alluded to, existentialism is very much part of the humanist sphere of culture, while analytic philosophy, however one defines it, takes its bearings and its problems from the scientific realm. I wish to argue that however dissimilar and unconnected they may seem to be, and, of course, the differences are vast, the two perspectives are nonetheless quite the same in one crucial respect: Their ultimate "solutions" rest upon the recognition, even if only implicitly, of the evolutionary, biological reality of human life. Nor, as I have tried to suggest in the preceding several pages, is this accidental. Exposure to the significance of history, of man's animal nature, and of modern culture produces a unique perspective which will produce unique styles of philosophizing. Existentialism and analytic philosophy are, respectively, the humanist and scientific versions of the uniquely contemporary philosophical perspective.

What I have just suggested will, I suspect, seem quite curious. In order to make sense of the argument we shall have to make summary statements of these rather complicated bodies of philosophical literature.

In an important way the writings of Hegel lie at the root of both existentialism and analytic philosophy, because both can properly be understood as in large part reactions against the Hegelian synthesis. Hegel, at the beginning of the nineteenth century, saw the bifurcation of Western culture and attempted to put it back together again by taking a detached observer position with respect to history. Hegel, of course, could not have known the significance of geological and evolutionary time—he was several decades too early—but because he was a Western man he could and did grasp the importance of developmental change in human history. The Hegelian overview required the perspective of the detached observer. Geoffroy-Dechaume distinguishes the Western from the Chinese view: "Time and space, those two systems of measurement without which man could not be assessed, have never been considered in China as abstractions, detached from particular events and places. The Chinese do not, as we do, conceive of some continuous flux in the course of which events occur, nor of some void into which places are inserted. Time, for the Chinese, is and always has been an aggregate of eras, seasons, periods; space, a complex of realms, climates, directions." * But Hegel saw human history as the rational manifestation, in time, of God's potentiality through the agency of man, as "the march of God on earth."

Hegel's view, then, is what we might call "anticipatory-contemporary" in that he sees, partially at least, the significance of time through the detached observer perspective. Rather the same thing, incidentally, may be said of Marx. Marx is a power in our time precisely because he makes revolutionary social change a "scientific" doctrine at just the point when the Western discoveries of time and of the detached observer perspective are spreading around the globe. The proof of this is the obvious fact that Marxism is relevant and successful neither in the modern West nor in genuinely primitive cultures but in fringe areas where modern culture is just beginning.**

For both Hegel and Marx any particular individual is but a moment or an instance in the movement of history. What both underemphasize is the adaptive, evolutionary, biological reality of each human life, and it was left, therefore, to existentialism and analytic philosophy, albeit in very different ways, to stress this element in the contemporary perspective.†

---

* François Geoffroy-Dechaume, *China Looks at the World* (New York: Pantheon Books, 1967), p. 172.

** Cf. Adam Ulam, *The Unfinished Revolution* (New York: Random House, 1959).

† Freud, of course, picks up the same point but from a less explicitly philosophical angle. The twentieth-century merger of psychoanalysis with existentialism by Ludwig Binswanger and others is, thus, not surprising.

Harvard philosopher Stanley Cavell, after having commented upon the differences and indeed the considerable animosities between the two views, suggests: "Yet both are modern philosophies; both are, by intention and in feeling, revolutionary departures from traditional philosophy. That is, perhaps, a characteristic of philosophy generally: every departure believes itself to be escaping from an empty, hateful past, and to be setting the mind at last on the right road. Yet it is striking that the terms 'analytical' and 'existential' were initially coined to purify philosophy of the identical fool's gold in its tradition—the tendency to issue in speculative systems [particularly that of Hegel]. The discovery of analytical philosophy is that such systems make statements which are meaningless or useless; the discovery of existentialism is that such systems make life meaningless." *

4

What does it mean to be an existentialist, or to think "existentially"? It is possible, of course, to become involved in a very elaborate discussion in pursuit of an answer to this question. If, however, we focus on the central stream of existentialist thinking, it is not unreasonable to argue that one who accepts Sartre's famous maxim "existence precedes essence" and makes it the basis of his position thinks existentially. Heidegger has rejected this particular statement, but nonetheless if one is careful about explaining it, the general sense of existentialism can be indicated.

In an important respect the maxim is the contradiction of a line of thought implicit in philosophy from Aristotle to Hegel. For both can reasonably be understood as assenting to a statement that "essence precedes existence." The essence of man was conceived as a sort of category or slot in the nature of the universe into which each man fits more or less well. Thus, the philosopher's task is to discover the nature of man so that each man can use it as a guide. This is "is-to-ought" reasoning in its classic form. But existentialism from its contemporary perspective rejects this elemental notion.

"Existence precedes essence" might have been announced as proper doctrine at various stages in the long history of philosophy. A medieval nominalist, for example, might have said it. He would have understood it to mean that particular objects, for example, animals, are prior to the universal concept "animal," which is in fact simply their common name.

---

* Stanley Cavell, "Existentialism and Analytical Philosophy," *Daedalus,* Summer, 1964, p. 948. Reprinted by permission of *Daedalus,* Journal of the American Academy of Arts and Sciences, Boston, Mass., Summer, 1964, "Existentialism and Analytical Philosophy."

Similarly, a modern empiricist, even an analytic philosopher, might say it and mean by it that perceived facts are prior to the general laws that explain them. "When the existentialist says it, however, he means something more and different. He is thinking of *man's* essence and existence. And he is saying that when a person tries to concentrate on the universal essence of man instead of on the living and poignant existence directly exemplified in himself, he is turning away from human reality instead of growing toward a sound understanding of it. By this route one will inevitably find man's essence in his faculty of reason, and will employ merely rational thinking in the endeavor to understand it." *

Thus, the existentialist thinker like Nietzsche (who in this respect can be considered an existentialist) sees clearly the tension between truth, the product of reason, and life, which can properly be understood as the product of evolution.

> A man . . . is a full person, not just a cognitive mind, and it requires all the resources of a full person to understand him. To win these resources one must face the fact of one's own existence, with its emotional involvements and its fateful possibilities of weal and woe. Everyone has been thrown into the turbulent current of life, and whether he is aware of it or not this predicament is the determinative factor in all that he does—including his apparently rational thinking. Until we recognize, in this setting, that existence precedes essence, one cannot hope to understand man or the deeper realities of human experience.**

The brooding Norseman, Søren Kierkegaard, who is correctly regarded as the founder of existentialist thinking, began as a Hegelian but rejected Hegel's rational, universal system in favor of the reality of individual life. Kierkegaard retained Hegel's sense of time, of dialectical growth, but saw it in the context of individual development through creative choice. He saw the human condition from, so to speak, the "inside" as a matter of free, animal adaptation—not, however, as instinctive response to environment, but in the uniquely human context of conscious choice in the face of environment and animal instinct. In this way he and the existentialists who follow him were profoundly biological. Kierkegaard discovered from the inside what Darwin and Freud discovered from the outside.

One more point about Kierkegaard and the existentialists is extremely important for our discussion. Once Kierkegaard had grasped the tension between life and truth, between living and reasoning, he saw that reasoning in the ordinary sense of deducing and inducing was inadequate to the

* E. A. Burtt, *In Search of Philosophic Understanding* (New York: The New American Library, 1965), pp. 77–78.
** *In Search of Philosophic Understanding*, p. 78.

problem of choosing, of deciding, in actual life. One cannot *know* in a scientific or superscientific (that is, traditional metaphysical) sense, but one must live, and to live, one must choose. The standard of choice he found difficult to communicate. It must, he suggests, be communicated indirectly. Thus it is that existentialists often eschew philosophical demonstration in the style of solving algebra problems in favor of a literary, poetic form.* The test is one of "authenticity"—the questions and the answers must be "authentic" to the matter of living itself. Interpretations, demonstrations, and objections directed at matters of choice, of ethics, of religion—in short, of life—derived from the realm of abstract reason are inauthentic, are artificial, and thus distort the problems at hand.

5

While Kierkegaard's understanding of man's biological condition through the notion of individual existence is clear enough, the point with respect to analytic philosophy is not so easy to see. It will be particularly difficult for someone who thinks that analytic philosophy means Logical Positivism, and this seems to be the interpretation of a substantial majority of nonphilosophers, particularly social scientists. For them the following remarks by one of America's outstanding analytic philosophers, Stanley Cavell of Harvard, will seem very strange.

> Kierkegaard's diagnosis of our illusion, our illness, in the *Concluding Unscientific Postscript,* is that we have lost the capacity for subjectivity, for inwardness, and therewith the capacity for Christianity. We live in an Objective Age, an Age of Knowledge, and we have stopped *living* our lives in favor of knowing them. Wittgenstein's diagnosis is that we have, in part because of our illusions about language, fixed or forced ideas of the way things must be, and will not *look and see* how they are. Kierkegaard finds us trying to escape our existence and our history; Wittgenstein finds us trying to escape the limits of human forms of language and forms of life. In Kierkegaard's descriptions, we live in the universal rather than in our particularity; in Wittgenstein's we crave generality instead of accepting the concrete.**

In order to make sense of these remarks we must realize that Cavell is talking about the mature Wittgenstein, the Wittgenstein of the *Philosophical Investigations,* and not about the Wittgenstein of the *Tractatus,* the inspirer of Logical Positivism. As I pointed out in the preceding chapter

---

* Cf., Thomas Hanna, *The Lyrical Existentialists* (New York: Atheneum, 1962).
** Cavell, "Existentialism and Analytical Philosophy," *Daedalus,* pp. 958–959. (Italics in original.)

Wittgenstein in his early life came to philosophy from logic and the foundations of mathematics and, in so doing, clearly stated the implications of the visually isolated style of thinking which had developed in the West since 1500. He focused his entire argument, as I indicated earlier, around the written proposition and the way "like a scale applied to reality" it pictured the world of facts. The power of his argument stimulated the Logical Positivists and aroused their fierce attack upon the humanist philosophical tradition, but, as he later realized, the clarity of his argument also brought this style of thinking to its dead end. The great misfortune is that in the last several decades social scientists have attached themselves to this dead end and have proclaimed it far and wide, supposing it to be the revolutionary ultimate discovery of truth.

Throughout his life Wittgenstein approached philosophy through language. While his view changes radically in his later writings, the focus on language is a constant. After having pushed the relationship between the written language and the world to its logical end, he saw its inadequacy as a representation of genuine human circumstances. A new world opens when he directs himself to the oral, living language. Compare this pivotal and typical passage from the *Philosophical Investigations* with the sections of the *Tractatus* quoted earlier:

> But how many kinds of sentence are there? Say assertion, question, and command?—There are *countless* kinds: countless different kinds of use of what we call "symbols," "words," "sentences." And this multiplicity is not something fixed, given once for all; but new types of language, new language-games, as we may say, come into existence, and others become obsolete and get forgotten. (We can get a *rough picture* of this from the changes in mathematics.)
>
> Here the term "language-game" is meant to bring into prominence the fact that the *speaking* of language is part of an activity, or of a form of life.
>
> Review the multiplicity of language-games in the following examples, and in others:
> Giving orders, and obeying them—
> Describing the appearance of an object or giving its measurements—
> Constructing an object from a description (a drawing)—
> Reporting an event—
> Speculating about an event—
> Forming and testing a hypothesis—
> Presenting the results of an experiment in tables and diagrams—
> Making up a story; and reading it—
> Play-acting—
> Singing catches—
> Guessing riddles—
> Making a joke; telling it—
> Solving a problem in practical arithmetic—

Translating from one language to another—
Asking, thanking, cursing, greeting, praying.

—It is interesting to compare the multiplicity of the tools in language and of ways they are used, the multiplicity of kinds of word and sentence, with what logicians have said about the structure of language. (Including the author of the *Tractatus Logico-Philosophicus*.) *

Thus, Ludwig Wittgenstein, uniquely perhaps in the history of philosophy, rejects his early sharply rational view, and, significantly, he rejects it in the name of life, the forms of life revealed in the living language. His change of view is well illustrated by two anecdotes in which the sources of his inspiration are revealed. The first concerns the formation of the central idea of the *Tractatus,* the picture theory of meaning.. It is related by the Finnish philosopher Georg Henrik von Wright:

Wittgenstein told me how the idea of language as a *picture* of reality occurred to him. He was in a trench on the East front [Wittgenstein served with the Austrian army in World War I], reading a magazine in which there was a schematic picture depicting the possible sequence of events in an automobile accident. The picture there served as a proposition; that is, as a description of a possible state of affairs. It had this function owing to a correspondence between the parts of the picture and things in reality. It now occurred to Wittgenstein that one might reverse the analogy and say that a *proposition* serves as a *picture,* by virtue of a similar correspondence between its parts and the world. The way in which the parts of the proposition are combined—the *structure* of the proposition—depicts a possible combination of elements in reality, a possible state of affairs.**

The exclusively visual, printed-page character of the perception around which Wittgenstein built his doctrine of the *Tractatus* could not be clearer than in this illustration. It is the detached-observer-to-reality relationship reduced to its essence. I think, moreover, that there is a sense in which Wittgenstein understood this fact. As his friend Norman Malcolm reports, "Also he told me once that he really thought that in the *Tractatus* he had provided a perfected account of a view that is the *only* alternative to the viewpoint of his later work." †

It is Malcolm who relates the second incident which marks the beginning to Wittgenstein's *philosophical* encounter with the language of life, thus opening the door to the teaching of the *Philosophical Investigations*:

* Ludwig Wittgenstein, *Philosophical Investigations* (New York: Macmillan, 1953), sec. 23, pp. 11ᵉ–12ᵉ. (Italics in original.)
** From *Ludwig Wittgenstein: A Memoir* by Norman Malcolm, with an Introduction by G. H. von Wright, published by the Oxford University Press, pp. 7–8. (Italics in original.)
† *Ludwig Wittgenstein: A Memoir,* p. 69. (Italics in original.)

Wittgenstein and P. Sraffa, a lecturer in economics at Cambridge, argued together a great deal over the ideas of the *Tractatus*. One day (they were riding, I think, on a train) when Wittgenstein was insisting that a proposition and that which it describes must have the same "logical form," the same "logical multiplicity," Sraffa made a gesture, familiar to Neapolitans as meaning something like disgust or contempt, of brushing the underneath of his chin with an outward sweep of the finger-tips of one hand. And he asked: "What is the logical form of *that*?" Sraffa's example produced in Wittgenstein the feeling that there was an absurdity in the insistence that a proposition and what it describes must have the same "form." This broke the hold on him of the conception that a proposition must literally be a "picture" of the reality it describes.*

One cannot help wondering about the factors that were involved in Wittgenstein's coming back to ordinary language from his discussions of the logically purified language in the *Tractatus*. While the foregoing story illustrates the change very nicely, it can hardly be all that is involved. Probably, as I have suggested, his feeling that he had solved all philosophical problems *from that particular point of view* in the *Tractatus* plays a considerable role. What he realized was that the print-dominant, visually-isolated point of view was not the only point of view or even, indeed, the most important one. If there is a dominant theme in the *Investigations* it is, as I have earlier indicated, that a form of language is a form of life. Somehow the notion that language is a living thing and not simply "a scale applied to reality" came to him in a powerful way.

I am not enough of a McLuhanite to be utterly serious about noting the change in the dominant medium of communication that occurred between the writings of 1918 and those published after World War II. Reading McLuhan's discussion of "hot" and "cool" media—that is, "hot" as sharply defined, exclusively visual media such as print, and "cool" as absorbing, multisense media such as radio and television—makes it impossible, however, for me to resist calling your attention to Malcolm's description of one of Wittgenstein's regular habits during the late thirties and early forties. McLuhan argues that the electronic, "cool" media draw styles of thinking and feeling back toward the multisense milieu of oral culture, and remember that Malcolm at the time of his writing could have known nothing of McLuhan's views.

As members of the class began to move their chairs out of the room he might look imploringly at a friend and say in a low tone, "Could you go to a flick?" On the way to the cinema Wittgenstein would buy a bun or cold pork pie and munch it while he watched the film. He insisted on sitting in the very first row of seats, so that the screen would occupy his entire field

* *Ludwig Wittgenstein: A Memoir*, p. 69.

of vision, and his mind would be turned away from the thoughts of the lecture and his feelings of revulsion. Once he whispered to me "This is like a shower bath!" His observation of the film was not relaxed or detached. He leaned tensely forward in his seat and rarely took his eyes off the screen. He hardly ever uttered comments on the episodes of the film and did not like his companion to do so. He wished to become totally absorbed in the film no matter how trivial or artificial it was, in order to free his mind temporarily from the philosophical thoughts that tortured and exhausted him. He liked American films and detested English ones. He was inclined to think that there *could not* be a decent English film. This was connected with a great distaste he had for English culture and habits in general. He was fond of the film stars Carmen Miranda and Betty Hutton. Before he came to visit me in America he demanded in jest that I should introduce him to Miss Hutton.*

So this was Wittgenstein. By all accounts his personality was extraordinarily intense. He lived philosophy and through his latter analysis of language he philosophized life. What I suggest is that from his position in the scientific culture he drove through the rigidities of Logical Positivism into the biological reality of human life as expressed in ordinary language. My point could not be more sharply made than in the words of Wittgenstein's most illustrious successor in the English analysis of ordinary language, Oxford philosopher J. L. Austin: "Our common stock of words embodies all the distinctions men have found worth drawing, and the connections they have found worth marking, in the life-time of many generations." These words, Austin continues, "are likely to be much more numerous, more sound since they have stood up to the long test of the survival of the fittest, and more subtle . . . than any that you or I are likely to think up in our armchairs of an afternoon." **

The idea that ordinary language could be a product of man's cumulative experience, a "survival of the fittest" is common sense in the twentieth century. It is a mode or angle of understanding readily available and even *obvious* in our time in a way that would not have been obvious at all three or four hundred years ago. And this is my point. It is not that Kierkegaard, Wittgenstein, or Austin precisely saw that they were reasoning biologically, that they were taking into account development through time and the evolutionary connection with nature, but the solutions to their particular

---

* *Ludwig Wittgenstein: A Memoir,* pp. 27–28. Italics in original. McLuhan, of course, classifies the motion picture as a "hot" medium, but by his own standards it was not hot the way Wittgenstein treated it. See also Paul Engelmann, *Letters from Ludwig Wittgenstein With a Memoir* (Oxford: Basil Blackwell, 1967), esp. pp. 90–93.

** J. L. Austin, "A Plea for Excuses." In *Philosophical Papers,* edited by J. O. Urmson and G. J. Warnock (London: Oxford University Press, 1961), Chapter 6. See p. 130.

problems arose from an understanding of the world where these perspectives were available, or even commonplace.

The existentialists and the ordinary language philosophers both understand that problems of understanding and problems of choosing are problems of living and not a matter of finding an abstract proof in the way that one would for a problem in geometry. Cavell, in comparing Wittgenstein and Kierkegaard, does not explicitly state his interpretation in biological terms, but his analysis nonetheless helps to make my case:

> I might summarize what I have been saying in somewhat the following way: Wittgenstein and Kierkegaard take seriously the fact that we begin our lives as children; what we need is to be shown a path, and helped to take steps; and as we grow, something is gained and something is lost. What helps at one stage does not help at another; what serves as an explanation at one stage is not serviceable—we could say, it is not intelligible—at another. In grown-up philosophy, the problems we have *remain* answerable only through growth, not through explanation or definitions. And grown-ups give themselves as many useless or fraudulent explanations as they give children. What we must hope for is not that at some stage we will possess all explanations, but that at some stage we will need none. And the task remains to discover what we need. Wittgenstein puts it this way: We impose a requirement (for example, of perfection, or certainty, or finality) which fails to satisfy our real need. Kierkegaard could also have said that, as he also suggests that we impose such requirements upon ourselves for just that reason, to avoid seeing what our lives really depend on.*

Against this background we shall now turn our attention in an explicit way to matters of political philosophy. As Cavell suggests, what Wittgenstein and Kierkegaard both recognize is that traditional philosophizing has led us astray with respect to matters of human choice whether at a moral, religious, or political level. We have been taught to attempt to resolve choices by looking for proofs logically parallel to the solutions to mathematical or scientific problems. This, say both Wittgenstein and Kierkegaard, is misleading; it is logically erroneous. We must turn not to the logic of proof, but to what might be called the logic of experience or in the present context what we can call biological logic.

Consider for a moment what is involved in learning to drive an automobile and, in particular, learning to make a left turn in traffic. Anyone presented with this problem and with a recurrence of similar but never identical situations adopts through experience a rule or a set of rules of behavior. "When to go" and "when to wait" builds up in the driver's mind and in his "second nature" over time. If he is a professional driver, say, a taxi driver, he will probably put a finer edge on the decision and he may be bolder given what making the turn quickly means to him. The

* Cavell, "Existentialism and Analytical Philosophy," *Daedalus,* p. 970.

rule varies with cultural context. An American driver will notice that he must adjust to different parts of the country. If his driving experience is wider, he will realize that making a left turn in Madison, Wisconsin is a very different thing from making a left turn in Toronto, Rome, or Paris—or, heaven protect him, Manila.

Notice that if one is pressed to justify his left turn rule, he could, by wriggling and twisting the situation, force it into something that sounds like a deductive or inductive proof of the rule. The important thing to see, however, is that no matter how clever we may be about stating the matter deductively or inductively, this is not in fact how we learn to make left turns. What we do is not irrational, but neither is it inductive or deductive. The process has a logic and it is the logic of experience. This logic of life is what Kierkegaard and Wittgenstein saw. This is life *as it is* and not as we would intellectualize it. As Malcolm reports, "When I once quoted to him a remark of Kierkegaard's to this effect: 'How can it be that Christ does not exist, since I know that he has saved me?' Wittgenstein exclaimed: 'You see! It isn't a question of proving anything!' " *

6

The recognition of the priority of life over truth which I have been trying to describe in the last few pages can produce a new perspective on political philosophy. The first steps in this direction have already been taken and I should like to say something about them. While I have argued that existentialism and analytic philosophy come in the final analysis to something like the same thing, it is nonetheless true that their original foci and, thus, passageways to solution are quite different. One is essentially humanistic, the other scientific.

The central problem of political philosophy has always been one of reconciling the individual with the whole. One may philosophize about the whole from the perspective of the individual or vice versa. The arguments which I should like to discuss are, first, one arising from the existentialist perspective which, not at all surprisingly, takes the individual as point of departure. It is advanced by Albert Camus. The second takes its inspiration from the scientific culture and from analytic philosophy and is written from the perspective of the political system as a whole. This argument was presented a few years ago—I hope you will forgive me—by a political scientist named Thorson.** We turn first to Camus.

Camus's early philosophizing was not directed primarily at social or political matters; his concern was, rather, highly personal. He opens *The*

* Malcolm, *Ludwig Wittgenstein: A Memoir,* p. 71.
** I discover that this parallel is well discussed elsewhere. See Fred H. Willhoite, Jr., *Beyond Nihilism* (Baton Rouge: Louisiana State University Press, 1968), esp. pp. 196–201.

Myth of Sisyphus with a paragraph which shows his problem as personal and emotional as well as logical:

> There is but one truly serious philosophical problem, and that is suicide. Judging whether life is or is not worth living amounts to answering the fundamental question of philosophy. All the rest—whether or not the world has three dimensions, whether the mind has nine or twelve categories—comes afterwards. These are games; one must first answer. And if it is true, as Nietzsche claims, that a philosopher, to deserve our respect, must preach by example, you can appreciate the importance of that reply, for it will precede the definitive act. These are facts the heart can feel; yet they call for careful study before they become clear to the intellect.*

It is by no means unusual to pose "whether or not life is worth living" as a philosophical question, but it is surely a bit peculiar to focus on the question in terms of the problem of suicide. Camus's real interest is not in the phenomenon of suicide, but rather in the validity of a nihilist philosophical position. For Camus the modern world is above all marked off by "the death of God" in Nietzsche's sense. Traditional theological and philosophical standards which gave meaning to the world and to the life of man are no longer intellectually available to modern man or, to be more precise, what would have to be called, from Camus's point of view, "clear-thinking modern man." The modern world has become, in a way which Camus catalogues at length in *The Myth,* a nihilist world. The nihilist position is purported to follow from man's increased awareness of the irrationality of the universe. When man confronts the world lucidly, he can find no meaning; he finds instead what Camus, like the existentialists before him, calls "the absurd." If "the absurd" in truth leads to a nihilist position, it must answer the question "is life worth living?" in the negative and, thus, suicide becomes the proper logical alternative. It is in this sense that suicide becomes the primary philosophical problem—"The subject of this essay is precisely this relationship between the absurd and suicide, the exact degree to which suicide is a solution to the absurd." **

The notion of "the absurd" first makes its appearance at the level of feeling; it is a matter of emotional response. From a philosophical point of view, it has, as Camus suggests, "a ridiculous beginning." Man becomes abruptly aware of the absence of sense in the fundamental rules and habits which guide his life. "It happens," says Camus, "that the stage sets collapse."

> Rising, streetcar, four hours in the office or the factory, meal, streetcar, four hours of work, meal, sleep, and Monday Tuesday Wednesday Thursday

---

* From *The Myth of Sisyphus,* by Albert Camus, trans. by Justin O'Brien. © Copyright 1955 by Alfred A. Knopf, Inc. Reprinted by permission of the publisher.
** *The Myth of Sisyphus,* p. 5.

Friday and Saturday according to the same rhythm—this path is easily followed most of the time. But one day the "why" arises and everything begins in weariness tinged with amazement. "Begins"—this is important. Weariness comes at the end of the acts of a mechanical life, but at the same time it inaugurates the impulse of consciousness and provokes what follows. What follows is the gradual return into the chain or it is the definitive awakening. At the end of the awakening comes in time, the consequence: suicide or recovery.*

The condition of being "well adjusted," of being at home in the world, is for Camus a consequence of human mental construction either conscious and deliberate or unconscious and habitual. The absurd arises when it becomes clear that the arrangement of compatibility between man and the world is not truth but construction.

> At the heart of all beauty lies something inhuman, and these hills, the softness of the sky, the outline of these trees at this very minute lose the illusory meaning with which we had clothed them, henceforth more remote than a lost paradise. The primitive hostility of the world rises up to face us across millennia. For a second we cease to understand it because for centuries we have understood in it solely the images and designs that we had attributed to it beforehand, because henceforth we lack the power to make use of that artifice. The world evades us because it becomes itself again. That stage scenery masked by habit becomes again what it is. It withdraws at a distance from us.**

Camus does not for a moment pretend that there is anything startling or original about these observations; they have, of course, often been noted by a wide variety of thinkers. But Camus, as we suggested earlier, is inclined to relate the emotional to the logical, and thus he here introduces an acute epistemological analysis by a description of an emotional reaction. Similarly, his conclusion involves both logic and a commitment to act in a certain way.

Camus's position which I have, perhaps a bit inappropriately, called epistemological is one of the inadequacy and fallibility of the human mind. "Of whom and of what indeed can I say: 'I know that.' This heart within me I can feel, and I judge that it exists. This world I can touch, and I likewise judge that it exists. There ends all my knowledge, and the rest is construction." † His reaction to the teachings of natural science is particularly acute:

> You describe the world to me and you teach me to classify it. You enumerate its laws and in my thirst for knowledge I admit that they are true. You

---

* *The Myth of Sisyphus*, p. 10.
** *The Myth of Sisyphus*, p. 11.
† *The Myth of Sisyphus*, p. 14.

take apart its mechanism and my hope increases. At the final stage you teach me that this wondrous and multicolored universe can be reduced to the atom and that the atom itself can be reduced to the electron. All this is good and I wait for you to continue. But you tell me of an invisible planetary system in which electrons gravitate around a nucleus. You explain this world to me with an image. I realize then that you have been reduced to poetry: I shall never know. . . . I realize that if through science I can seize phenomena and enumerate them, I cannot, for all that, apprehend the world. Were I to trace its entire relief with my finger, I should not know any more. And you give me the choice between a description that is sure but that teaches me nothing and hypotheses that claim to teach me but that are not sure.*

Camus translates a hazy emotional response into a sharply analytical description of the context in which all men operate or, in language closer to the tone of Camus's remarks, of "the human condition." This context or condition has three parts. There is man, there is the world, and there is that which relates man and the world, that which at once makes the world inscrutable and the human mind fallible. This "absurd" situation is for Camus simply given; it is the one truth that is certain; it is the place to start philosophical analysis. Moreover,

The immediate consequence is also a rule of method. The odd trinity brought to light in this way is certainly not a startling discovery. But it resembles the data of experience in that it is both infinitely simple and infinitely complicated. Its first distinguishing feature in this regard is that it cannot be divided. To destroy one of its terms is to destroy the whole. There can be no absurd outside the human mind. Thus, like everything else, the absurd ends with death. But there can be no absurd outside this world either. And it is by this elementary criterion that I judge the notion of the absurd to be essential and consider that it can stand as the first of my truths. The rule of method alluded to above appears here. If I judge that a thing is true, I must preserve it. If I attempt to solve a problem, at least I must not by that very solution conjure away one of the terms of the problem. For me the sole datum is the absurd.**

Camus suggests three ways of dealing with the absurd: suicide, hope, or living with it. Camus rejects the alternative of suicide because it is an escape from, rather than a solution to, the problem of the absurd; suicide eliminates the problem, it does not solve it. The alternative of hope comes in effect to the same thing; it is, as Camus very acutely suggests, philosophical suicide. Hope is found in an alleged solution to the absurd which lies beyond knowledge. It may be "God" or "history" or "reason," but such a solution again does not solve the problem, it eliminates it by

* *The Myth of Sisyphus*, p. 15.
** *The Myth of Sisyphus*, p. 23.

arguments for which there is insufficient evidence. Here Camus is sharply critical of those existentialists who urge a leap into the irrational. The absurd being the only truth, the essence of the human situation, the only proper alternative is to live with it. He later cogently summarizes his argument in the introductory passages of *The Rebel*:

> The final conclusion of absurdist reasoning is, in fact, the repudiation of suicide and the acceptance of the desperate encounter between human inquiry and the silence of the universe. Suicide would mean the end of this encounter, and the absurdist reasoning considers that it could not consent to this without negating its own premises. According to absurdist reasoning, such a solution would be the equivalent of flight or deliverance. But it is obvious that absurdism hereby admits that human life is the only necessary good since it is precisely life that makes this encounter possible and since, without life, the absurdist wager would have no basis.*

At core Camus's argument is an affirmation of the value of human life, and there is surely nothing very startling about this conclusion. What is important is the *way* in which he argues for this conclusion. He speaks quite consciously as modern man for whom the traditional arguments are unavailable. His discussion is neither idle nor academic; one almost gets the feeling that he is personally trying to decide whether to commit suicide or not. To be able to live without appeal to the transcendental or the suprarational—this is his objective. He tries to show that so-called nihilist premises do not lead to nihilist conclusions, and he speaks therefore directly to those for whom the nihilist premises appear sound.

So far this is but a solution to a personal problem—whether my own life is worth living in the light of the absurd. But it cannot remain merely personal, for to refuse death in the name of the absurd is one thing, but to *live* in the name of the absurd is another. Living is a social problem, and Camus must therefore turn to social philosophy. His efforts are at first quite unsure; it is almost as if he experiments in literature. Can society be properly regulated by an abstract set of absolute moral and legal norms? This, it seems to me, is certainly one, if not *the,* important question of his celebrated novel, *The Stranger*. It should be noted that the critics have not read it in these terms, but it seems to me one of the most persuasive indictments of the notion of objective liability in criminal law ever written. The protagonist, a man who could not be more ordinary, finds himself in a set of circumstances which culminate in his killing a man. From the reader's standpoint it is clearly not a case of murder, but from inside the story and particularly from the point of view of the law and its application, it is murder, and the protagonist is sentenced to death. Camus sharply portrays the contrast between life as it is and the objective norms which seek to regulate it.

---

* Albert Camus, *The Rebel* (New York: Alfred A. Knopf, Inc., 1956), p. 6.

Can society, on the contrary, be governed by the absurd? Camus's first major play, *Caligula,* is directed at this question. The emperor Caligula starts with the absurdist premise "since there is no right or wrong, everything is permitted" and proceeds to rule by mere personal whim. Camus's dramatic and philosophical suggestion at the end of the play is that everything is not permitted and that the absurd carries within itself what Camus calls "limits."

The problem of the consequences of the absurd becomes for Camus more directly and immediately political with the Nazi occupation. His four remarkable short essays which he called "Letters to a German Friend" are written in this context. The potential consequences of nihilism had become actual, and for Camus the problem was therefore all the sharper. His reaction here is more from the heart than from the mind, but it suggests what is to come. He addresses his "German friend" in part as follows:

> For a long time we both thought that this world had no ultimate meaning and that consequently we were cheated. I still think so in a way. But I came to different conclusions from the ones you used to talk about, which, for so many years now, you have been trying to introduce into history. I tell myself now that if I had really followed your reasoning, I ought to approve what you are doing. And this is so serious that I must stop and consider it, during this summer night so full of promises for us and of threats for you.
>
> You never believed in the meaning of this world, and you therefore deduced the idea that everything was equivalent and that good and evil could be defined according to one's wishes. You supposed that in the absence of any human or divine code the only values were those of the animal world—in other words, violence and cunning. Hence you concluded that man was negligible and that his soul could be killed, that in the maddest of histories the only pursuit for the individual was the adventure of power and his only morality, the realism of conquests. And, to tell the truth, I, believing I thought as you did, saw no valid argument to answer you except a fierce love of justice which, after all, seemed to me as unreasonable as the most sudden passion.
>
> Where lay the difference? Simply that you readily accepted despair and I never yielded to it. Simply that you saw the injustice of our condition to the point of being willing to add to it, whereas it seemed to me that man must exalt justice in order to fight against eternal injustice, create happiness in order to protest against the universe of unhappiness. Because you turned your despair into intoxication, because you freed yourself from it by making a principle of it, you were willing to destroy man's works and to fight him in order to add to his basic misery. Meanwhile, refusing to accept that despair and that tortured world, I merely wanted men to rediscover their solidarity in order to wage war against their revolting fate.
>
> As you see, from the same principle we derived quite different codes, because along the way you gave up the lucid view and considered it more convenient (you would have said a matter of indifference) for another to

do your thinking for you and for millions of Germans. Because you were tired of fighting heaven, you relaxed in that exhausting adventure in which you had to mutilate souls and destroy the world. In short, you chose injustice and sided with the gods. Your logic was merely apparent.

I, on the contrary, chose justice in order to remain faithful to the world. I continue to believe that this world has no ultimate meaning. But I know that something in it has a meaning and that is man, because he is the only creature to insist on having one. This world has at least the truth of man, and our task is to provide its justifications against fate itself. And it has no justification but man; hence he must be saved if we want to save the idea we have of life. With your scornful smile you will ask me: what do you mean by saving man? And with all my being I shout to you that I mean not mutilating him and yet giving a chance to the justice that man alone can conceive.*

But what kind of rational defense is to be offered for what Camus here calls "justice"? Clearly no Platonic argument can be relevant for a man who holds "the absurd" as the only truth. Camus had suggested in *The Myth of Sisyphus* that one of the consequences of the absurd was what he called "revolt." This notion of revolt became the principle vehicle of his developing political philosophy. He experimented with it in his plays *The State of Siege* and *The Just Assassins* and in his complex novel, *The Plague,* and everywhere he seemed to reach with difficulty and often with something less than clarity a notion that "revolt" implied "limit" or, in other words, some standard of positive value.

In his essay *The Rebel* he takes on this problem squarely. He settles here, again in a way foreign to any Anglo-American philosopher, on the problem of murder. Where suicide, the personal matter, was formerly the problem, he argues, now murder, the social question, is the problem. Camus's argument is complex and difficult, and any summary is likely to be rather shallow, but let me with this warning attempt to suggest in a few words what Camus uses many to show. If the absurd leads to the denial of suicide, it affirms the value of life. Such an affirmation means not only refusing to take one's own life, but also refusing to allow anyone else to take it and, similarly, refusing to take the life of another. From the beginning, then, this refusal or this "revolt" implies a set of limits, but as a matter of history the problem is by no means this simple. Revolt as it occurs historically is revolt against something, that is, revolt against some particular oppression or set of oppressions. Recognition of the absurd leads immediately to the denial of traditional values and to revolt against the historical embodiment of those values, that is, the existing authority, in the name of personal value. Thus, concretely it leads to murder.

* From *Resistance, Rebellion and Death,* by Albert Camus, trans. by Justin O'Brien. © Copyright 1960 by Alfred A. Knopf, Inc. Reprinted by permission of the publisher.

Revolt (or, more precisely, "perverted revolt" which Camus calls "rebellion") destroys the old values because they have no value from the absurdist perspective, but as a matter of history seeks to replace them with new values. Thus, murder begins in the name either of the absurd or of new values, and it continues in the name of preserving those new values. The bulk of *The Rebel* is an attempt to describe the perversion of revolt by the creation of the new values of "reason" (in the case of the French Revolution) and "history" (in the case of the Marxist revolution). In short it is an attempt to show the major "revolts" of recent years as "perverted revolt." Revolt in the name of socialism has become, in one of Camus's most memorable phrases, "socialism of the gallows."

But, Camus insists, revolt in the name of life cannot consistently end in murder. This really is the force of *The Rebel*: Revolt when carefully analyzed contains its own limits, not limits which are easily specified, but limits nonetheless. And all this without appeal to the transcendental.

In 1957 Camus published an essay entitled "Reflections on the Guillotine." While the bulk of it is a spine-tingling description of capital punishment in France, it contains also his final and perhaps most succinct statement in support of the rights of man against the state, a subject which had been in many ways his central concern for twenty years. The problem of the modern world is not, says Camus, the sort of murder which the guillotine was invented to punish; that is, "private" murder, a crime against society; but "public" murder, the act of society against the individual. The murders committed by the Nazi state alone are hundreds of times more numerous than all of the "private" murders of the twentieth century. His argument against state-imposed death, what he calls rational murder, hinges on the very fact that he speaks as a modern man. It was once possible, he argues, to reasonably support the death penalty on Christian grounds, for man's judgment even if it were in error could always be superseded by the final judgment of God. But for Camus, modern man has no such excuse:

> For the majority of Europeans, faith is lost. And with it, the justifications faith provided in the domain of punishment. But the majority of Europeans also reject the State idolatry that aimed to take the place of faith. Henceforth in mid-course, both certain and uncertain, having made up our minds never to submit and never to oppress, we should admit at one and the same time our hope and our ignorance, we should refuse absolute law and the irreparable judgment. We know enough to say that this or that major criminal deserves hard labor for life. But we don't know enough to decree that he be shorn of his future—in other words, of the chance we all have of making amends. Because of what I have just said, in the unified Europe of the future the solemn abolition of the death penalty ought to be the first article of the European Code we all hope for.*

* *Resistance, Rebellion and Death*, pp. 229–230.

Here, a bit more explicitly than elsewhere, Camus voices a defense of men against the state, a defense which is in some ways strikingly similar to an argument which might be advanced by a pragmatist. Society has no right to kill in the absence of absolute certainty, first, that it possesses a metaphysical or religious right to do so, and second, that there is no possible doubt in a particular case that there is no error of legal judgment. Given the essentially limited character of human knowledge—both in general, philosophical matters, and in the practical details of evidence—neither of these conditions can ever be realized. But Camus's argument is not merely pragmatic; it is a pragmatism enriched by the existential arguments reviewed earlier. Man is not simply the fallible cognitive machine—a sort of inadequately programmed computer—he is also a being aware of imminent death for whom his own fallibility reveals the precious quality of life. It is significant to say that, because we can never be certain that we are completely right, we can never be justified in performing actions as irremediable as taking away a man's life, but it is more significant to affirm simultaneously the value of that life.

7

I began my first writing in political theory in the summer of 1959 and finished what I came to call *The Logic of Democracy* * in the spring of 1960. Putting myself in the position of reviewing my own book is in some ways quite uncomfortable, but it has the virtue that I pretty well know what I intended to say and how the argument came to take the form that it did. After a few years have passed, books have a way of detaching themselves from their authors and taking on a sort of life of their own. Because I have a fairly complete recollection of what went through the author's mind, I may indeed be uniquely positioned to describe this document which is no longer part of me.

My considered conclusion is that the argument of *The Logic of Democracy* is essentially correct, but the author did not really understand what he was doing. In a curious way the present book, although it is written some eight years later can serve as an introduction to *The Logic of Democracy,* because, I think, the argument there presented makes considerably more sense against the background of the present pages. It is fair to say, I suppose, that my suggestion of the last several pages to the effect that twentieth-century men respond to the peculiar intellectual milieu of the twentieth century, even though they may not explicitly know that this is what they are doing, derives in considerable measure from my

* New York: Holt, Rinehart and Winston, 1962.

own experience. *The Logic of Democracy* is a biological book even though I did not know it when I wrote it.

In 1959 I was emerging from a distinctly American program of higher education. As a political scientist *cum* political theorist I was bothered by the problem of the justification of the democratic order in the face of an intellectual style—relativist, positivist, behavioralist—which insisted that no justification beyond culturally conditioned personal preference was possible. Arnold Brecht's "tragedy of twentieth-century political science," the apparent inability to justify democracy as opposed to the various forms of totalitarianism, seemed a genuine tragedy to me. Moreover, it seemed to me curious that democracy as a matter of fact was a roaring success, "preferred" and vigorously defended by millions of apparently rational human beings, even though no truly persuasive intellectual argument could apparently be made for it. One had to conclude in the face of this situation that either these millions were not rational, or that there was something wrong with our criteria for and standards of intellectual argument. My judgment was that democracy did indeed make sense, but that somehow our rules of intellectual activity had gone astray. *The Logic of Democracy* became a critique of these rules.

The core of the argument was based on the logical parallel between the *justification* of scientific method and the *justification* of democratic method. One cannot, it seemed to me, "prove" scientific method to be "true" by deductive demonstration from some grand principle immanent in the nature of things. Indeed, it makes no sense to talk this way. How would one go about "proving" that any "way of doing something" was "true"? Moreover, it would be ridiculous to say that by employing scientific method the inquirer would inevitably emerge with the "right" answer. How would he know that it was "right"? By the same token it would make no sense to stack up instances of the application of scientific method and attempt to arrive at some sort of inductive generalization with respect to them. What could the generalization possibly prove? That scientific method produced "right" answers? Again, how would you know that they were "right"? Is there some test external to scientific method in terms of which it can be evaluated? If so, what is it? The mind of God? But, if you knew the mind of God, why would you need scientific method?

Having, thus, pointed out the impossibility of justifying scientific method by deduction or induction, I turned to what I called the logic of recommendation. If one wants to recommend the adoption of scientific method, how is it possible to justify such a recommendation? Following Charles Sanders Peirce, I argued that the recognition of the impossibility of knowing absolute truth through deduction, induction, or direct inspection leads to a categorical statement of a rule for the investigator: "Do not block the way of inquiry." This rule, this recommendation, was, I argued, a statement of the attitude which developed from the recognition of the eternal impossibility of knowing the absolute truth about anything.

If I had it to do over again, I think that I would put the same point in a slightly different way. I would talk primarily about the logic of experience rather than the logic of recommendation, because it now seems to me that the logic of experience is prior and at the same time more easily communicated. J. Robert Oppenheimer once observed that learning to be a scientist is a matter of apprenticeship, and indeed it is. Grasping the scientific attitude is a matter of experience. Learning the "rules" of science is like learning to be a quarterback or learning how to make left turns in traffic. And in an important way learning to be a scientist is like Kierkegaard's sense of learning to be a Christian. What the recommendation does is to show us a path, help us take the first steps, and the recommendation of the path is justified because the route has been explored. Its pitfalls have been spotted and its dead ends discovered. A set of procedures develops that prevents "blocking the way of inquiry," and when the recommendation is put into words it reflects the logic of experience and is not a matter of deduction or induction.

My political argument was parallel to the scientific one. Stressing the similarity—although, of course, recognizing the differences—between "making decisions about the way the world is" in science and "making decisions about what a society should do" in politics, I put the general recommendation for politics in the form, "Do not block the possibility of change with respect to social goals." The justification for this rule consisted in the recognition that no one could *know* the proper goals for any society and could not, therefore, be justified in imposing his goals on others. The problem for the scientific community as for the political community, once these recommendations are accepted, is to find ways of institutionalizing the attitudes expressed by the recommendations.

The scientists do it by enforcing norms of proper scientific procedure: Examine the facts, erect tentative hypotheses, test the hypotheses by experiment, submit the results publicly for testing by others, and so forth. Respecting such norms prevents blocking the way of inquiry.

In politics a parallel process of institutionalization is necessary: Open the door to all opinions with respect to goals, make the decisions responsive to all involved, and so on. Institutionalizing the recommendation "Do not block the possibility of change with respect to social goals," I suggested, amounts to adopting a political method embodying the principles of popular sovereignty, political equality, protection of individual and minority rights, and majority rule. It makes sense, therefore, to adopt democracy in the spirit of the famous remark by Sir Winston Churchill, "Democracy is the worst form of government except for all other forms."

By the same token scientific method is the worst form of investigation (it is tedious, cumbersome, exhausting) except for all other forms.

The preceding paragraphs, of course, represent only the barest summary of the core of the argument. For a more elaborate discussion I must refer you to *The Logic of Democracy* itself. The foregoing continues to seem sensible to me as far as it goes. I now want to discuss the respects in which the author of *The Logic of Democracy* did not quite know what he was doing. This can best be done by raising the most common objection to the thesis, an objection raised by a number of critics and one which I regularly hear from my students.

You begin, they say, by making a strong argument against justification by deduction. Classical deductive arguments for democracy proceed from broad metaphysical premises such as, "All men are created in the image of God" or "All men are created equal." From premises so broad anything or nothing can follow. "All men are created in the image of God" in the hands of one who wants to support democracy is nothing more than a very elaborate way of stating a preference for democracy. It is not an external, independently established truth from which democracy follows, but rather part of a very large tautology which simply states the speaker's preference for democracy. It is a pseudoproof and therefore not a justification at all.

This is indeed the argument that I make. Most of the critics seem to follow it and in most cases to accept it. But then they turn it back on me. Your argument, they say, that the recognition of human fallibility leads to the justification of scientific method and democratic method is in fact nothing more than a covert piece of metaphysical deduction and thus falls prey to the arguments against deduction. The major premise, they suggest, is "All men are fallible" and the argument, therefore, is no better than the one that starts with "All men are created in the image of God."

This argument is in fact quite specious, but against the argument, *as it is put* in *The Logic of Democracy* it has a certain force. The reason is that in *The Logic of Democracy* I relied explicitly and, in an explicit sense, exclusively on a static conception of logic. There exists, I suggested, deductive logic, inductive logic, and a rather vague category which I called the logic of recommendation. What I did not altogether understand was that what I called the logic of recommendation was not a static set of rules for "truth getting" like deduction or induction, but the time-oriented, experience-oriented logic of life. I had read some of what Wittgenstein, Kierkegaard, Nietzsche, and Camus *said,* but I did not really unde· ;tand what they *meant.* Indeed, as I have previously suggested, I am not c rtain that they fully understood the implications of what they said. The fact is that scientific method and democratic method are justified as a matter of life and not of truth. This is what I was trying to say in *The Logic of Democracy* but I did not know quite how to do it.

Let me restate the argument diagrammatically:

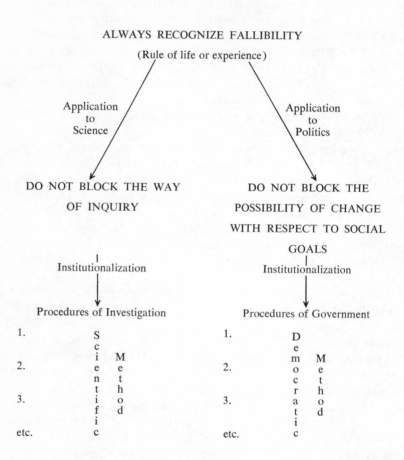

ALWAYS RECOGNIZE FALLIBILITY

(Rule of life or experience)

Application to Science

Application to Politics

DO NOT BLOCK THE WAY OF INQUIRY

DO NOT BLOCK THE POSSIBILITY OF CHANGE WITH RESPECT TO SOCIAL GOALS

Institutionalization

Institutionalization

Procedures of Investigation

Procedures of Government

1.   Scientific Method   1.   Democratic Method

2.   2.

3.   3.

etc.   etc.

This is a picture of the argument as I presented it in *The Logic of Democracy*. Put this way it is easy to see how the argument has a deductive look about it. The soundness of my case turns on the way in which the rule "Always recognize fallibility" is established. Does it amount to saying "All men are fallible" or not? If it does, then "All men are fallible" is just another metaphysical premise and the argument is just another example of the loose, metaphysical deduction which can be so thoroughly demolished.

But in the spirit of Wittgenstein we must ask not How can we intellectualize it? but How *is* it as a matter of ordinary life? Again, remember the fact that modern science and modern mass democracy are the products of Western culture since 1500. Their methods and styles are justified as a matter of time-oriented life experience. My diagram might better be put in this way:

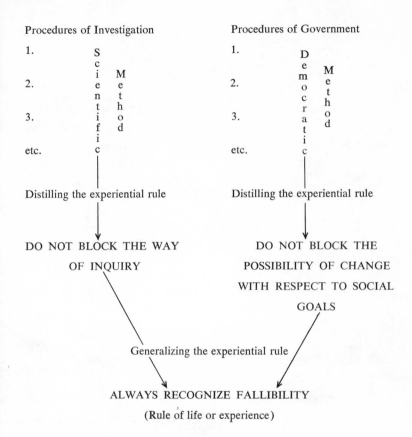

Procedures of Investigation

1.

2.

3.

etc.

Scientific Method

Distilling the experiential rule

DO NOT BLOCK THE WAY OF INQUIRY

Procedures of Government

1.

2.

3.

etc.

Democratic Method

Distilling the experiential rule

DO NOT BLOCK THE POSSIBILITY OF CHANGE WITH RESPECT TO SOCIAL GOALS

Generalizing the experiential rule

ALWAYS RECOGNIZE FALLIBILITY

(Rule of life or experience)

I wrote in *The Logic of Democracy*: "Let us face the fact that in one sense the justification of fallibilism as the overriding 'must' of science is circular. It is generally supposed that calling a piece of reasoning 'circular' is the most devastating attack that can be leveled against it; but it is a mistake to suppose that all circles are vicious. What we are contending is that we are rationally obligated to behave as if we are limited with respect to knowledge of matters of fact because we *are* so limited. The examination of our own tools of analysis reveals these limitations. Knowledge of our instruments shows us at once what we can do and what we cannot do." *

Thus, the full and proper diagram of the argument should look like this:

* Page 123.

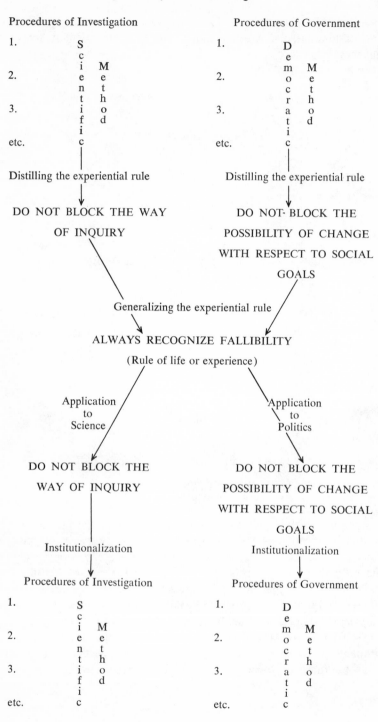

Justifying scientific method or democratic method is a matter of "showing why it makes sense to commit yourself to" scientific method or democratic method. The key to understanding the argument diagramed above is seeing that it strives to be "authentic" with respect to such choices. Becoming committed to a way of doing something is a matter of *individual* choice. Becoming *rationally* committed to such a choice is a matter of learning from one's own experience or from the experience of someone else. It is for just this reason that existentialist thinkers from Kierkegaard to Camus have attempted to give instruction on choice by revealing experience in literary form. In *Caligula* or *The Just Assassins* Camus attempts to do more than to assert "that there are limits;" he attempts to *show* that there are limits.

When Camus's man decides not to commit suicide, he affirms the value of *his* life not by reference to something general about man or about the nature of things, but by virtue of *his own* confrontation with the absurd. He could not *prove* that all men are valuable by examining them in some scientific sense. Such an argument is not only impossible—"the world is absurd" in Camus's terms; "deduction, induction, or direct inspection cannot establish absolute truth" in Peirce's language—it is also irrelevant. It is irrelevant in the sense of "inauthentic." Commitment to the value of life comes not from an external demonstration but from the inner push of life itself. In the same way, for Kierkegaard, one cannot *be* a Christian by virtue of some external state of affairs—having been born a Northern European—one must *become* a Christian by dint of a lifetime of personal choices.

As "All men are valuable" is impossible to prove, irrelevant, and inauthentic to the man contemplating suicide, so also is "All men are fallible" to the man committing himself to scientific or democratic method. Note carefully the tension between "authentic choice" and "intellectualized or rationalized choice" in what I have just said. It is easy, you may say to me, to imagine a man talking himself out of suicide with a phrase like "All men are valuable." Indeed, but see that Camus speaks to the question of what *really counts"*—that *I* am valuable—and not about the way in which that commitment may be expressed, articulated, or intellectualized. If one takes "All men are valuable" as his reason, justification of this reason involves nothing short of examining all men and showing that they are valuable. Surely this is impossible—how would you go about finding the property "valuable"? But when this argument fails, *suicide does not follow*! Thus the argument is irrelevant and inauthentic to the actual business of choosing.

In just the same way when it comes to choosing (either in the sense of "setting up" or "committing oneself to") a set of rules for investigating reality or a set of rules for deciding what a society should do, any kind of "All men are X" statement including "All men are fallible" is irrelevant and inauthentic to the personal commitment itself. Another way of making the point is to say that there is nothing at all abstract about being committed to democracy. Being committed to democracy means letting the guy talk even when you hate the son-of-a-bitch and think he is 400 percent wrong, but it also means refusing to allow him to destroy that highest political achievement of cultural evolution, the rules of the democratic order. Expressing this kind of commitment as a matter of logic was the task of *The Logic of Democracy*.

8

In reviewing these two lines of political argument—the one existentialist and individually oriented, the other based on the analysis of the logic of language and system oriented—I have been attempting to show that both are conscious, even if only implicitly, of the evolutionary perspective. Both reflect the logic of life against the background of contemporary Western culture. Neither, however, pushes through to the truly contemporary synthesis—for neither reflects evolution become conscious of itself. It is just this perspective that the book you are now reading has attempted to state. By *biopolitics*—the politics of life—I mean politics understood by man as evolution becomes conscious of itself.

The contemporary problem for political understanding is one of getting itself clear about the character of the contemporary world. I choose the term "political understanding" carefully because if the present analysis shows anything, it shows that the radical distinction between political science and political philosophy, like the broader distinction between "is" and "ought," is a product of the print culture of the last four hundred years. When one faces the question of politics in an evolutionary way, the distinction, at least as a matter of philosophy if not as a matter of the practice of the individual investigator, simply evaporates.

Karl Marx in his limited nineteenth-century way understood this point. His political science and his political philosophy were one and the same, and he made no attempt to conceal this fact. One of the aspects of the contemporary world that political understanding must get itself clear about is the significance of Marxism. While some of us suspect—rightly, I think—that Marxism is on the wane, none can deny its extraordinary and in some ways unprecedented success. Yet we are confused about the reasons for its success. We know that it spread by force and violence, but force and violence are nothing new. The question remains: Why should force and violence be so successful tied to this particular doctrine? The answer, I think, is really quite simple, but one must have the proper perspective to see it. The leap of the West to a new cultural level after 1500 has rapidly accelerated the fact of social change, first in the West itself and then in all those areas touched by the West. The fact of change through time—the "four-dimensionality" of human existence which was always present but was previously invisible—suddenly and increasingly became a primary fact of life. And Marxism, almost alone among accounts of social and political life, is a theory of society and politics in four dimensions. Social life is four-dimensional and so is Marxism and this is its power.

Again the history of science can instruct us. We began this long discussion by contrasting the Newtonian universal-generalization paradigm of understanding with the Darwinian evolutionary-developmental paradigm of understanding. A similar contrast—indeed perhaps the same contrast—lies at the core of the current political division of the world. By the eighteenth century the Western world had begun to move quickly enough so that the sensitive could recognize it. By the nineteenth century, particularly on the Continent of Europe itself, in France and in Germany, the sense of movement jarred the passengers with a violence that only a fool could have missed.

The Englishmen of the seventeenth and eighteenth centuries, together with certain of their Continental friends, notably Montesquieu, saw the fact of flux in human affairs. In the style of their intellectual guide, Sir Isaac Newton, they sought to build a structure based on universal laws that would be strong enough to weather whatever storm arose. This stance is typically expressed in two metaphors, one architectural and the other contractual.* In the first case the architect of the state takes the detached observer position outside political life itself. It is his duty not only to understand men as they are, but to shape them into a new political edifice that will stand the test of time. In the second case the idea of social contract expresses the notion of creating man-made but durable products called civil society and government. Harrington, for example, describes "props and scaffolds" for the building of a "Constitution which stands by it selfe" from thence forward. In the rhetoric of Madison, Hamilton, and Adams the constitution was an "edifice," "erected" upon "foundations," with "pillors," "interior," and "superstructure."

The image of the hydroelectric dam built upon a swiftly flowly river which we used earlier to describe the theory of the political system is also an apt way to describe the constitutional structure of Harrington and Madison. It is therefore not surprising that Easton's theory should have descriptive value for the American polity. After all, Madison and the others shared with Easton the Newtonian paradigm and the edifice which they built was informed by it.

* I am indebted to Professor Kirk Thompson of Reed College for his discussion of these matters in his paper "Constitutional Theory and Political Action" delivered at the 1967 meeting of the American Political Science Association.

Marx, the nineteenth-century German—we ought not to forget—dedicated *Das Kapital* to Darwin.* Darwin shared with Herder, Hegel, and Marx the evolutionary, developmental paradigm. Where the dead hand of the past obstructed the accelerating social and economic change of the nineteenth and twentieth centuries, for Marx and his many followers the prospect of erecting an intricately engineered dam over a river by that time in raging flood seemed both impossible and irrelevant. In this context Marx provided a political science and a political philosophy which was essentially a time-oriented theory of change, what I call here a four-dimensional theory. This conception stands in sharp, paradigmatic contrast with the static, enduring processing machine of Hobbes, Locke, Harrington, Montesquieu, and Madison. In the main it was only the English speakers (and significantly, the English *readers*), who had built their processing machine early, who could for the most part ignore the four-dimensional conception of Marx. For the Russians and the Chinese it became profoundly relevant and it could not be ignored by the French, the Germans, and the Italians.

But, as in science, paradigms of understanding in human affairs have a way of becoming dogmas. They become conceptual *Gestalten* and their propagation involves myth and slogan. This is probably especially so where the paradigm has been written down and its enunciation therefore takes on the semimagical authority of an ancient text. We have only to reflect upon the American intellectual and moral struggle, concentrated in the 1930s but by no means yet over, with the eighteenth-century conception of the natural right to private property.** The same point becomes clear from a consideration of the dogmatically inspired collectivization of agriculture by the Russians and the Chinese no matter what the cost in blood, misery, and, for that matter, in agricultural production itself.

Marx's four-dimensionality, though its essential orientation is correct, is nonetheless a brutal distortion. In its inception, as Camus has so brilliantly argued, it is a revolt in the name of life, but in its realization it is a regime of death. And Camus protests, bringing forward for all the Marxists to see the existential—or, if you will, the individual, developmental—reality of every human life without which the evolutionary movement of society as a whole is empty of truly human meaning. Camus shouts, "But there are limits!" in the face of the pious executioners of the socialism of the gallows. What Camus cannot readily see from his existentialist,

---

* For some reason or another this fact has lately become the object of scholarly attention. Marx didn't mean it and Darwin didn't like it, some say, but it is a fact nonetheless.

** Cf., Thurmond Arnold, *The Folklore of Capitalism* (New Haven: Yale University Press, 1937) and C. B. MacPherson, *The Real World of Democracy* (London: Oxford University Press, 1967).

personally-oriented perspective is just what the stodgy old eighteenth-century English constitutionalists saw with consummate clarity: that the limits to be effective must not only be felt, they must be institutionalized.

Insofar, however, as the constitutionalists based their institutions and their rules on a static conception of natural rights which allegedly attach to individuals like arms and legs, the understanding and application of these institutions and rules inevitably became strained and distorted as society became more complex and more urban. The last seventy years or so of American constitutional history has been an attempt to overcome the rigidities of the eighteenth-century argument.* Thus, it becomes necessary in the middle of the twentieth century to reexamine the grounds for these institutions and rules, not simply because the eighteenth-century argument is no longer persuasive, but also because the eighteenth-century conception (centered as it was in the notion of a natural right to private property in land) provides guidelines—and standards for the interpretation of guidelines—which are not appropriate to twentieth-century conditions. The notion of commitment to democratic institutions because they embody the recommendation "Do not block the possibility of change with respect to social goals" opens the way to adjustment not only to the problems of the twentieth century but to those of the twenty-first century as well. What this kind of argument seeks to express is a four-dimensional as opposed to a hydroelectric dam conception of decision-making institutions and rules.

The justification for such a position is at its core best stated by Camus's affirmation of the value of individual life through the recognition of the elemental, existential—infrarational, if you will—challenging of the absurd. Camus affirms further that the recognition of this elemental human challenge implies the refusal to allow one's life to be taken and the refusal, furthermore, to take the life of another. What Camus's perspective makes difficult is the precise prescribing of the rules that will make this attitude operative in real societies. Here the careful analysis—arising from the scientific culture—of architectonic political theorizing comes to fit Camus's vital humanism like the other half of a jigsaw puzzle.

In the final analysis political problems are solved or "muddled through" in the streets, on the battlefields, in the offices of governments, and in the hearts and minds of men. The record, however, is profoundly clear that human ingenuity can move the problems and the solutions in this way or that. We can hope that the ingenuity of Camus's statement of limits can penetrate the attitude so well put by Yevgeny Yevtushenko:

> He who is conceived in a cage yearns
> for the cage.

* Cf., Chapter 9 of *The Logic of Democracy*.

> With horror I understood that I love
> That cage where they hide me behind
>   a paling
> And that animal farm, my home land.

Thus, when we, from an open-ended, evolutionary perspective, merge an institutional structure based on the recommendation of leaving open the possibility of change with respect to social goals with an existential affirmation of the existence of limits, we may say that we have solved a problem at the intellectual level. The problem of social order is one of obligating men to do what they really want to do on personal grounds. Institutionalizing the limits implied in personal revolt is the solution, at the intellectual level, to that problem. Doing more is a matter of the ingenuity of politics itself.

Perspective is, however, all important. Our perspective is four-dimensional and we must stand with Père Teilhard: "One might well become impatient or lose heart at the sight of so many minds (and not mediocre ones either) remaining today still closed to the idea of evolution, if the whole of history were not there to pledge to us that a truth once seen, even by a single mind, always ends up by imposing itself on the totality of human consciousness. For many, evolution is still only transformism, and transformism is only an old Darwinian hypothesis as local and as dated as Laplace's conception of the solar system or Wegener's Theory of Continental Drift. Blind indeed are those who do not see the sweep of a movement whose orbit infinitely transcends the natural sciences and has successively invaded and conquered the surrounding territory—chemistry, physics, sociology, and even mathematics and the history of religions. One after the other all the fields of human knowledge have been shaken and carried away by the same underwater current in the direction of the study of some *development*. Is evolution a theory, a system, or a hypothesis? It is much more: It is a general condition to which all theories, all hypotheses, all systems must bow and which they must satisfy henceforward if they are to be thinkable and true. Evolution is a light illuminating all facts, a curve that all lines must follow." *

* *The Phenomenon of Man,* London: Collins, pp. 218–219.

Index